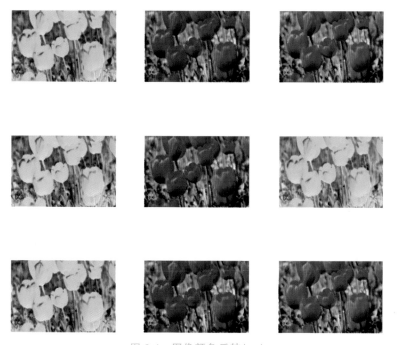

图 3-4 图像颜色反转(一)

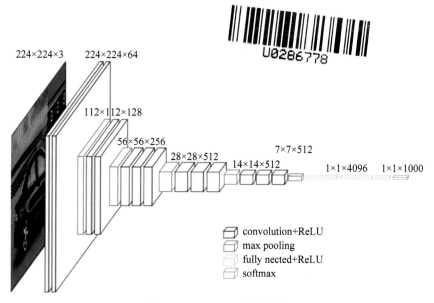

图 3-13 VGGNet 的网络结构

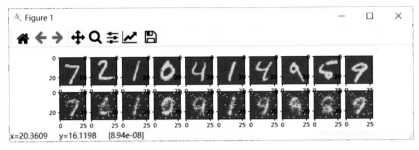

图 7-5 测试效果图

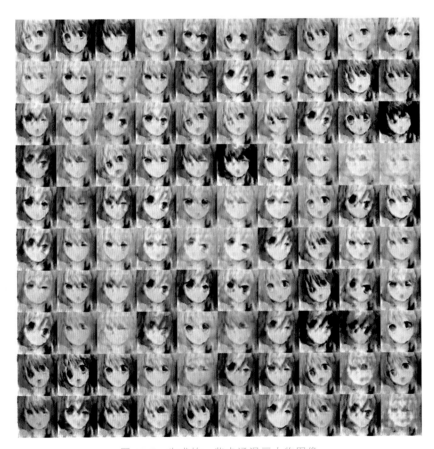

图 10-5 生成的一些卡通漫画人物图像

全国高等学校计算机教育研究会"十四五"规划教材

实战深度学习高阶
——运用TensorFlow

张智超 邓劲生 尹晓晴 编著

清华大学出版社
北京

内容简介

本书是一本介绍深度学习的实践教材,采用的深度学习工具为TensorFlow。本书主要内容包括深度学习环境搭建,数据处理与模型加载,图像处理,图像增强识别实战和两阶段目标检测实战,单阶段目标检测,实战文本识别和图像生成,机器创作,情感分类、翻译、对话,GAN及其变体的创作,强化学习与迁移学习等。降低了深度学习的门槛。

本书案例丰富新颖,涉猎领域广泛,代码分类清晰,实用性和可操作性强,可作为高校人工智能相关专业以及相关培训机构的案例实战教材,也可以作为深度学习智能应用及算法工程师强化训练的参考书。

本书封面贴有清华大学出版社防伪标签,无标签者不得销售。

版权所有,侵权必究。举报:010-62782989,beiqinquan@tup.tsinghua.edu.cn。

图书在版编目(CIP)数据

实战深度学习高阶:运用TensorFlow/张智超,邓劲生,尹晓晴编著. —北京:清华大学出版社,2022.4
ISBN 978-7-302-60057-2

Ⅰ.①实… Ⅱ.①张… ②邓… ③尹… Ⅲ.①机器学习 Ⅳ.①TP181

中国版本图书馆 CIP 数据核字(2022)第 016551 号

责任编辑:白立军
封面设计:傅瑞学
责任校对:焦丽丽
责任印制:沈 露

出版发行:清华大学出版社
  网　　址:http://www.tup.com.cn,http://www.wqbook.com
  地　　址:北京清华大学学研大厦 A 座　　邮　编:100084
  社 总 机:010-83470000
  投稿与读者服务:010-62776969,c-service@tup.tsinghua.edu.cn
  质量反馈:010-62772015,zhiliang@tup.tsinghua.edu.cn
  课件下载:http://www.tup.com.cn,010-83470236
印 装 者:三河市龙大印装有限公司
经　　销:全国新华书店
开　　本:185mm×260mm　　印　张:16.5　　彩 插:1　　字　数:405 千字
版　　次:2022 年 4 月第 1 版　　　　　　　　　　　　印　次:2022 年 4 月第 1 次印刷
定　　价:59.80 元

产品编号:088980-01

前 言

总的来说,计算机视觉、文本处理、游戏对弈等深度学习最新前沿,最终目标就是使智能体能够像人一样看待和理解这个世界。本书通过介绍一个个经典案例,穿插讲解代码和其背后深度学习与其他经典算法的原理。与此同时,高度注重理论与实践的结合,充分讲解与注释代码。并且每个案例后面都给出了全面的安装步骤,以及相关的实验截图。

本书采用的深度学习工具为 TensorFlow。它是最出名的开源工具,使深度学习的门槛大大降低,不管是人工智能专家,还是对深度学习不很在行的开发人员,都可以轻易利用它开发出 AI 程序,其 2.x 版本还极大地简化了代码的复杂性。

本书的主要目的就是对最前沿的几个深度学习领域进行汇总。因此,为编写此书,作者查阅了大量论文和相关文献。本书共分为四部分。

第一部分介绍基础,包括第 1 章和第 2 章。第 1 章介绍了深度学习环境搭建。主要围绕深度学习的基本特点和规律,以及如何搭建相关的 CPU 与 GPU 版本的平台、安装 cuDNN 神经网络库、匹配 Python 与 Anaconda 版本等进行介绍。第 2 章介绍深度学习实战的基本流程。每次通过一个案例的方式介绍结构化数据、回归方法、过拟合、保存模型等最基本的深度学习实战处理基础知识和应用流程。

第二部分主要介绍计算机视觉领域的经典案例,包括第 3～6 章。第 3 章实现了对图像的基本操作,包括增强、分类、语义分割。第 4 章围绕增强图像的质量展开,包括去模糊、去噪声、提高分辨率实战案例。第 5 章实现了不同方法的目标检测。第 6 章介绍 YOLO、SSD 以及 DarkFlow,是单阶段检测的经典实战案例。

第三部分是关于文本处理与生成实战,包括第 7～9 章。第 7 章主要围绕手写体问题展开,包括神经网络、编码器、对抗样本生成、姿态生成等新的知识点。第 8 章进入文本创作实战,包括创作音乐、歌词、古诗。第 9 章主要对情感分类、翻译、对话进行实战。

第四部分主要是对抗生成网络与强化学习实战,包括第 10 章和第 11 章。其中,第 10 章介绍 GAN,围绕生成实战串联了模仿建筑风格、相似动物转换、人脸生成等项目。第 11 章介绍强化学习与迁移学习在小游戏中的应用。

本书配有全套源代码和资源。代码讲解与使用方法在每节中的代码解析和安装操作两个环节中进行了明确。

本书是《实战深度学习——原理、框架及应用》的进阶学习材料。最初起源于团队自身建设的能力提升所需。我们改编了一批当前热门的应用案例作为实战化操作练习,并准备了全套源代码、数据集和使用说明等学习资源。本书随着团队新生力量的增加而不断更新,多次被作为培训教材使用且反响良好。

  本书是跨域大数据智能分析与应用省级重点实验室团队协同努力的成果，由邓劲生和尹晓晴负责搭建整体框架并确定实战内容、组织验证应用和调度实施，张智超选取案例并撰写了大部分章节，陈怡、严少洁、喻庭昌参与了部分章节撰写，调试了全部代码并整理优化文字，熊炜林、王良、曹吉浩、孙睿豪等进行了核查验证及资源梳理，乔凤才、宋省身、赵涛、李勐等老师参与了文稿修改完善指导。部分内容来自于参考文献和网络资源转载，未能逐一溯源和说明引用，特在此对这些资源的作者表示感谢。

  由于深度学习正处于蓬勃发展之中，而作者的自身水平、理解能力、项目经验和表达能力有限，书中难免存在一些错误和不足之处，还望各位读者不吝赐教，也欢迎将本书选作教材的老师垂询和交流，联系邮箱是 bljdream@qq.com。

作 者
2022 年 1 月于砚瓦池

目　录

第1章　深度学习环境搭建 1
1.1　深度学习典型应用 1
1.1.1　计算机视觉 1
1.1.2　自然语言处理 3
1.1.3　强化学习 3
1.2　Anaconda 使用简介 4
1.2.1　Anaconda 的特点 4
1.2.2　Anaconda 的下载及安装 4
1.2.3　Python 库的导入与添加 5
1.2.4　conda 命令简介 7
1.3　TensorFlow 使用简介 7
1.3.1　TensorFlow 的特点 8
1.3.2　CPU 版 TensorFlow 环境搭建与调用 8
1.3.3　GPU 版 TensorFlow 环境搭建与调用 9
1.4　Jupyter Notebook 使用简介 15
1.4.1　安装 Jupyter Notebook 15
1.4.2　运行 Jupyter Notebook 15

第2章　基本流程——数据处理与模型加载 17
2.1　使用特征列队结构化数据进行分类：预测心脏病 17
2.1.1　背景原理 17
2.1.2　安装操作 18
2.1.3　代码解析 19
2.1.4　训练测试 23
2.2　回归方法：预测燃油效率 24
2.2.1　背景原理 24
2.2.2　安装操作 24
2.2.3　代码解析 24
2.2.4　训练测试 29
2.3　过拟合与欠拟合：不同模型选择对结果的影响 33
2.3.1　背景原理 33
2.3.2　安装操作 34

2.3.3　代码解析 …………………………………………………………… 34
　　　2.3.4　训练测试 …………………………………………………………… 39
　2.4　保存与加载预训练模型 ………………………………………………………… 42
　　　2.4.1　背景原理 …………………………………………………………… 42
　　　2.4.2　安装操作 …………………………………………………………… 42
　　　2.4.3　代码解析 …………………………………………………………… 43
　　　2.4.4　训练测试 …………………………………………………………… 47

第3章　图像处理：增广与分类 ……………………………………………………… 49
　3.1　数据增广之图像变换实战 ……………………………………………………… 49
　　　3.1.1　背景原理 …………………………………………………………… 49
　　　3.1.2　安装操作 …………………………………………………………… 49
　　　3.1.3　代码解析 …………………………………………………………… 50
　　　3.1.4　训练测试 …………………………………………………………… 56
　3.2　在 CIFAR10 上应用 VGGNet 实现图像分类 ………………………………… 60
　　　3.2.1　背景原理 …………………………………………………………… 60
　　　3.2.2　安装操作 …………………………………………………………… 60
　　　3.2.3　代码解析 …………………………………………………………… 61
　　　3.2.4　训练测试 …………………………………………………………… 62
　3.3　图像识别之猫狗分类 …………………………………………………………… 63
　　　3.3.1　背景原理 …………………………………………………………… 63
　　　3.3.2　安装操作 …………………………………………………………… 63
　　　3.3.3　代码解析 …………………………………………………………… 64
　　　3.3.4　训练测试 …………………………………………………………… 65

第4章　图像增强识别实战 …………………………………………………………… 67
　4.1　应用高阶神经网络提高图像分辨率 …………………………………………… 67
　　　4.1.1　背景原理 …………………………………………………………… 67
　　　4.1.2　安装操作 …………………………………………………………… 67
　　　4.1.3　代码解析 …………………………………………………………… 68
　　　4.1.4　训练测试 …………………………………………………………… 70
　4.2　长短期记忆机器人识别色彩 …………………………………………………… 72
　　　4.2.1　背景原理 …………………………………………………………… 72
　　　4.2.2　安装操作 …………………………………………………………… 72
　　　4.2.3　代码解析 …………………………………………………………… 73
　　　4.2.4　训练测试 …………………………………………………………… 75
　4.3　使用注意力机制给图像取标题 ………………………………………………… 75
　　　4.3.1　背景原理 …………………………………………………………… 75
　　　4.3.2　安装操作 …………………………………………………………… 76

		4.3.3 代码解析 ………………………………………………………… 76
		4.3.4 训练测试 ………………………………………………………… 82

第 5 章 两阶段目标检测实战 ………………………………………………… 86

- 5.1 基于 RPN 实现目标检测 …………………………………………………… 86
  - 5.1.1 背景原理 …………………………………………………………… 86
  - 5.1.2 安装操作 …………………………………………………………… 87
  - 5.1.3 代码解析 …………………………………………………………… 88
  - 5.1.4 训练测试 …………………………………………………………… 92
- 5.2 应用 MTCNN 实现人脸目标检测 ………………………………………… 92
  - 5.2.1 背景原理 …………………………………………………………… 92
  - 5.2.2 安装操作 …………………………………………………………… 93
  - 5.2.3 代码解析 …………………………………………………………… 93
  - 5.2.4 训练测试 …………………………………………………………… 95

第 6 章 单阶段目标检测 ……………………………………………………… 97

- 6.1 应用单阶段完成目标检测之 YOLOv4 …………………………………… 97
  - 6.1.1 背景原理 …………………………………………………………… 97
  - 6.1.2 安装操作 …………………………………………………………… 98
  - 6.1.3 代码解析 …………………………………………………………… 98
  - 6.1.4 训练测试 …………………………………………………………… 101
- 6.2 应用锚定框完成目标检测之 SSD ………………………………………… 102
  - 6.2.1 背景原理 …………………………………………………………… 102
  - 6.2.2 安装操作 …………………………………………………………… 103
  - 6.2.3 代码解析 …………………………………………………………… 103
  - 6.2.4 训练测试 …………………………………………………………… 111
- 6.3 基于视频流目标检测之 DarkFlow ……………………………………… 114
  - 6.3.1 背景原理 …………………………………………………………… 115
  - 6.3.2 安装操作 …………………………………………………………… 115
  - 6.3.3 代码解析 …………………………………………………………… 116
  - 6.3.4 训练测试 …………………………………………………………… 118
- 6.4 基于 ResNet 实现目标检测 ……………………………………………… 121
  - 6.4.1 背景原理 …………………………………………………………… 122
  - 6.4.2 安装操作 …………………………………………………………… 122
  - 6.4.3 代码解析 …………………………………………………………… 124
  - 6.4.4 训练测试 …………………………………………………………… 128

第 7 章 实战文本识别和图像生成 …………………………………………… 129

- 7.1 应用卷积神经网络识别手写体文本 ……………………………………… 129

  7.1.1 背景原理 ······ 129
  7.1.2 安装操作 ······ 130
  7.1.3 代码解析 ······ 131
  7.1.4 训练测试 ······ 132
 7.2 应用自动编码器识别手写体文本 ······ 133
  7.2.1 背景原理 ······ 133
  7.2.2 安装操作 ······ 133
  7.2.3 代码解析 ······ 134
  7.2.4 训练测试 ······ 136
 7.3 应用变分自动编码器重建服饰 ······ 136
  7.3.1 背景原理 ······ 136
  7.3.2 安装操作 ······ 137
  7.3.3 代码解析 ······ 137
  7.3.4 训练测试 ······ 140

第 8 章 词汇与音乐的星空——机器创作 ······ 141
 8.1 创作钢琴曲之 Music Transformer ······ 141
  8.1.1 背景原理 ······ 141
  8.1.2 安装操作 ······ 142
  8.1.3 代码解析 ······ 144
  8.1.4 训练测试 ······ 147
 8.2 应用循环神经网络创作歌词 ······ 147
  8.2.1 背景原理 ······ 147
  8.2.2 安装操作 ······ 148
  8.2.3 代码解析 ······ 148
  8.2.4 训练测试 ······ 153
 8.3 应用门控循环单元创作歌词 ······ 154
  8.3.1 背景原理 ······ 154
  8.3.2 安装操作 ······ 156
  8.3.3 代码解析 ······ 156
  8.3.4 训练测试 ······ 158
 8.4 应用长短期记忆创作歌词 ······ 159
  8.4.1 背景原理 ······ 159
  8.4.2 安装操作 ······ 161
  8.4.3 代码解析 ······ 161
  8.4.4 训练测试 ······ 162
 8.5 应用文本生成器创作古诗 ······ 163
  8.5.1 背景原理 ······ 164
  8.5.2 安装操作 ······ 164

　　　　8.5.3　代码解析 ·· 165
　　　　8.5.4　训练测试 ·· 168

第9章　情感分类、翻译、对话 ·· **172**
　9.1　使用双向循环神经网络对电影评论情感分类 ·· 172
　　　　9.1.1　背景原理 ·· 172
　　　　9.1.2　安装操作 ·· 173
　　　　9.1.3　代码解析 ·· 173
　　　　9.1.4　训练测试 ·· 175
　9.2　应用词嵌入计算文本相关性 ·· 175
　　　　9.2.1　背景原理 ·· 176
　　　　9.2.2　安装操作 ·· 176
　　　　9.2.3　代码解析 ·· 176
　　　　9.2.4　训练测试 ·· 184
　9.3　中英翻译机器人 ·· 185
　　　　9.3.1　背景原理 ·· 185
　　　　9.3.2　安装操作 ·· 186
　　　　9.3.3　代码解析 ·· 186
　　　　9.3.4　训练测试 ·· 192
　9.4　基于Seq2Seq中文聊天机器人实战 ·· 192
　　　　9.4.1　背景原理 ·· 192
　　　　9.4.2　安装操作 ·· 193
　　　　9.4.3　代码解析 ·· 193
　　　　9.4.4　训练测试 ·· 197

第10章　GAN及其变体的创作 ·· **199**
　10.1　GAN的原理与实战 ·· 199
　　　　10.1.1　背景原理 ·· 199
　　　　10.1.2　安装操作 ·· 199
　　　　10.1.3　代码解析 ·· 201
　　　　10.1.4　训练测试 ·· 206
　10.2　应用生成对抗网络Pix2Pix模仿欧式建筑风格 ·· 207
　　　　10.2.1　背景原理 ·· 207
　　　　10.2.2　安装操作 ·· 207
　　　　10.2.3　代码解析 ·· 208
　　　　10.2.4　训练测试 ·· 212
　10.3　应用循环对抗网络CycleGAN完成相似动物转换 ·· 212
　　　　10.3.1　背景原理 ·· 212
　　　　10.3.2　安装操作 ·· 213

10.3.3 代码解析 ·········· 213
10.3.4 训练测试 ·········· 216
10.4 WGAN-GP 人脸生成实战 ·········· 217
10.4.1 背景原理 ·········· 217
10.4.2 安装操作 ·········· 218
10.4.3 代码解析 ·········· 219
10.4.4 训练测试 ·········· 225

第 11 章 强化学习与迁移学习 ·········· **226**
11.1 强化学习之玩转 *Flappy Bird* ·········· 226
11.1.1 背景原理 ·········· 226
11.1.2 安装操作 ·········· 227
11.1.3 代码解析 ·········· 227
11.1.4 训练测试 ·········· 232
11.2 使用 TensorFlow Hub 实现迁移学习预测影评分类 ·········· 232
11.2.1 背景原理 ·········· 232
11.2.2 安装操作 ·········· 233
11.2.3 代码解析 ·········· 233
11.2.4 训练测试 ·········· 236
11.3 使用预训练的卷积神经网络绘制油画 ·········· 236
11.3.1 背景原理 ·········· 236
11.3.2 安装操作 ·········· 237
11.3.3 代码解析 ·········· 237
11.3.4 训练测试 ·········· 244

结语 ·········· **248**

参考文献 ·········· **250**

第 1 章

深度学习环境搭建

本章将对深度学习以及本书所用到的开发工具进行简要介绍。

1.1 深度学习典型应用

深度学习算法已经广泛应用到人们生活的角角落落,例如手机中的语音助手、汽车上的智能辅助驾驶、超市和食堂的人脸支付等。下面从计算机视觉、自然语言处理和强化学习 3 个领域入手,为大家介绍深度学习的一些主流应用。

1.1.1 计算机视觉

1. 图像识别

图像识别(Image Classification)是常见的分类问题。神经网络的输入为图像数据,输出为当前样本属于每个类别的概率,通常选取概率值最大的类别作为样本的预测类别。图像识别是最早成功应用深度学习的任务之一,经典的网络模型有 VGG 系列、Inception 系列、ResNet 系列等。

2. 目标检测

目标检测(Object Detection)是指通过算法自动检测出图像中常见物体的大致位置,通常用边界框(Bounding Box)表示,并分类出边界框中物体的类别信息,如图 1-1 所示。常见的目标检测算法有 RCNN、FastRCNN、FasterRCNN、MaskRCNN、SSD、YOLO 系列等。

图 1-1 目标检测效果图

3. 语义分割

语义分割(Semantic Segmentation)通过算法自动分割并识别出图像中的内容,可以将语义分割理解为每个像素点的分类问题,分析每个像素点属于物体的类别,如图 1-2 所示。常见的语义分割模型有 FCN、U-net、SegNet、DeepLab 系列等。

图 1-2 语义分割效果图

4. 视频理解

随着深度学习在 2D 图像的相关任务上取得较好的效果,具有时间维度信息的 3D 视频理解(Video Understanding)任务受到越来越多的关注。常见的视频理解任务有视频分类、行为检测、视频主题抽取等。常用的模型有 C3D、TSN、DOVF、TS_LSTM 等。

5. 图像生成

图像生成(Image Generation)通过学习真实图像的分布,并从学习到的分布中采样而获得逼真度较高的生成图像。目前主要的生成模型有 VAE 系列、GAN 系列等。其中,GAN 系列算法近年来取得了巨大的进展,最新 GAN 模型产生的图像样本达到了肉眼难辨真伪的效果,如图 1-3 和图 1-4 所示为 GAN 模型的生成图像。

图 1-3 自动生成的图像　　　　图 1-4 艺术风格迁移效果图

除了上述应用,深度学习还在其他方向上取得了不错的效果,例如艺术风格迁移、超分辨率、图像去噪/去雾、灰度图像着色等一系列非常实用酷炫的任务,限于篇幅,不再赘述。

1.1.2 自然语言处理

1. 机器翻译

过去的机器翻译(Machine Translation)算法通常是基于统计机器翻译模型,这也是2016年前Google翻译系统采用的技术。2016年11月,基于Seq2Seq模型上线了Google神经机器翻译系统(GNMT),首次实现了源语言到目标语言的直译技术,在多项任务上实现了50%~90%的效果提升。常用的机器翻译模型有Seq2Seq、BERT、GPT和GPT-2等,其中,OpenAI采用的GPT-2模型参数量高达15亿个,甚至在发布之初以技术安全考虑为由拒绝开源GPT-2模型。

2. 聊天机器人

聊天机器人(Chat-bot)也是自然语言处理的一项主流任务,通过机器自动与人类对话,对于人类的简单诉求提供满意的自动回复,提高客户服务效率和服务质量,常应用在咨询系统、娱乐系统和智能家居等场景中。

1.1.3 强化学习

1. 虚拟游戏

相对于真实环境,虚拟游戏(Virtual Games)平台既可以训练、测试强化学习算法,又可以避免无关干扰,同时也能将实验代价降到最低。目前常用的虚拟游戏平台有OpenAI Gym、OpenAI Universe、OpenAI Roboschool、Deep MindOpen Spiel、MuJoCo等,常用的强化学习算法有DQN、A3C、A2C、PPO等。在围棋领域,AlaphGo程序已经超越人类围棋专家;在*Dota2*和《星际争霸》游戏上,OpenAI和DeepMind开发的智能程序也在限制规则下战胜了职业队伍。

2. 机器人

在真实环境中,机器人(Robotics)的控制也取得了一定的进展。如UC Berkeley在机器人的Imitation Learning、Meta Learning、Few-shot Learning等方向取得了不少进展。美国波士顿动力公司在人工智能应用中取得喜人的成就,如图1-5所示为其制造的机器人在自动行走、多智能体协作等任务上表现良好。

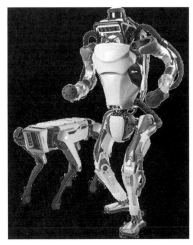

图1-5 波士顿动力公司的机器人

3. 自动驾驶

自动驾驶(Autonomous Driving)被认为是强化学习短期内能技术落地的一个应用方向,很多公司投入大量资源在自动驾驶上,如百度、Uber、Google公司的无人车等,其中,百度公司的无人巴士Apollo已经在北京、雄安、长沙等地展开试运营,图1-6为百度公司的

Apollo 自动驾驶汽车。

图 1-6　百度公司的 Apollo 自动驾驶汽车

1.2　Anaconda 使用简介

Anaconda 通过管理工具包、开发环境、Python 版本，大大简化了工作流程。Anaconda 不仅可以方便地安装、更新、卸载工具包，而且在安装时能自动安装相应的依赖包，同时还能使用不同的虚拟环境来隔离不同要求的项目[1]。

Anaconda 是一个非常方便的 Python 包管理和环境管理软件，它可以创建多个互不干扰的环境，分别运行不同版本的软件包，以达到兼容的目的。

1.2.1　Anaconda 的特点

Anaconda 有以下特点。
（1）包含众多流行的科学、数学、工程、数据分析的 Python 包。
（2）完全开源和免费。
（3）在使用中的额外加速和优化是收费的，但是对于学术用途可以申请免费的 License。
（4）全平台支持 Linux、Windows、macOS 等操作系统。

1.2.2　Anaconda 的下载及安装

Python 解释器是让 Python 语言编写的代码能够被执行的桥梁，是 Python 语言的核心。用户可以从 https://www.python.org/ 网站下载最新版本的 Python 解释器，和普通的应用软件一样安装完成后，就可以调用 python.exe 程序执行 Python 语言编写的源代码文件（*.py）。

这里选择安装集成了 Python 解释器和虚拟环境等一系列辅助功能的 Anaconda 软件，通过安装 Anaconda 软件，可以同时获得 Python 解释器、包管理、虚拟环境等一系列便捷功能。可以从网址 https://www.anaconda.com/distribution/#download-section 进入 Anaconda 下载页面，选择 Python 最新版本的下载链接，下载完成后即可进入安装程序。如图 1-7 所示，勾选 Add Anaconda to my PATH environment variable 复选框，这样可以通过命令行方式调用 Anaconda 的程序。整个安装流程持续 5～10min，具体时间依据计算机性能而定。

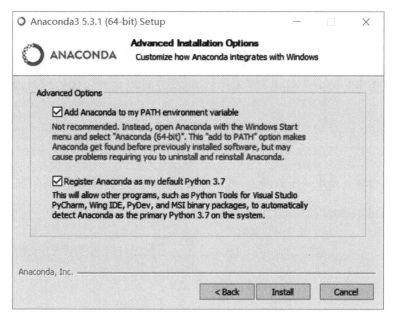

图 1-7　Anaconda 安装界面

安装完成后，在命令行输入 conda list 命令即可查看 Python 环境已安装的库。如果是新安装的 Python 环境，则列出的库都是 Anaconda 自带默认安装的软件库，如图 1-8 所示。如果 conda list 能够正常弹出一系列的库列表信息，说明 Anaconda 软件安装成功；如果 conda 命令不能被识别，则说明安装失败，需要重新安装。

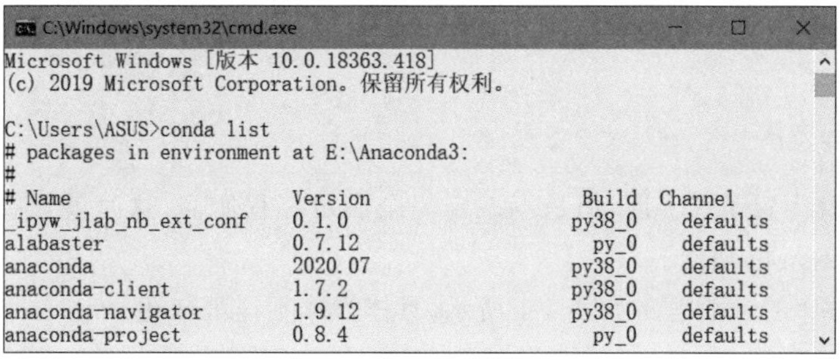

图 1-8　Anaconda 安装结果测试

1.2.3　Python 库的导入与添加

在搭建了 Python 基本平台 Anaconda 环境后，使用时一般不会将它全部功能都加载，但是当为了实现更多的功能时，则需要加载更多的库，必要时还可加载额外的第三方扩展库。

1. 库的导入

Python 内置了许多功能十分强悍的库，通过使用这些库可以实现非常复杂的功能。例

如内置的 math 库,可以提供正弦函数、指数等各种数学计算:

```
import math
math.sin(4)              #计算正弦值
math.exp(3)              #计算指数
math.pi                  #内置的圆周率常数
```

除了以上使用"import＋库名"导入库的方法,还可以在导入库的同时,给库起一个别名,例如:

```
import math as m
m.sin(3)                 #计算正弦
```

如果不需要导入库中的所有函数,可以特别指定导入函数的名字,例如:

```
from math import exp as e    #只导入 math 库中的 exp 函数,并起名为 e
e(1)                         #计算指数
sin(1)                       #此时 sin(1) 和 math.sin(1) 运行都会报错,因为并没有导入对应的包
```

2. 导入 future 特征

Python 的每个新版本都会增加一些新的功能,或者对原来的功能做一些改动。有些改动是不兼容旧版本的,也就是在当前版本运行正常的代码,到下一个版本运行可能就不正常了。例如 print,在 Python 2.x 中是关键字,用法主要是 print a;而在 Python 3.0 中,print 表示一个函数,用法为 print(a)。因此,为了保证程序的兼容性,在替换到高版本时,可以引入 future 特征。

```
from _future_ import print_fuction   #将 print 变成函数形式,即用 print(a)格式输出
from _future_ import division        # Python 3.0 中 5/2=2.5,5//2=2;Python 2.0 中 5/2=2
```

3. 添加第三方库

Python 自身内置了许多库,但是不一定完全满足编程所需,这时就需要导入一些第三方库来拓展 Python 的功能。

安装第三方库的一般思路如表 1-1 所示,具体操作不详细叙述。

表 1-1　安装第三方库的一般思路

思　路	特　点
下载源代码自行安装	安装灵活,但需要自行解决上级依赖问题
用 pip 安装	比较方便,自行解决上级依赖问题
用 easy_install 安装	比较方便,自行解决上级依赖问题,比 pip 稍弱
下载编译好的文件包	一般是 Windows 系统才提供现成的可执行文件包
系统自带的安装方式	Linux 或 macOS 的软件管理器自带某些库的安装方式

1.2.4 conda 命令简介

为了进一步帮助大家理解掌握 Anaconda 命令，加快环境配置优化更新，下面把本书相关的 conda 命令加以简要说明。更多详细的命令可查阅相关资料。

（1）在所在系统中安装 Anaconda。可以打开命令行输入 conda -V 检验是否安装以及当前 conda 的版本。

（2）conda 常用的命令。

conda list：查看安装了哪些包。

conda env list 或 conda info -e 或 conda info --envs：查看当前存在哪些虚拟环境。

conda update conda：检查更新当前 conda。

（3）创建 Python 虚拟环境。

可以使用 conda create -n *your_env_name* python=X.X（2.7、3.6、3.7 等）命令创建 Python 版本为 X.X、名字为 *your_env_name* 的虚拟环境。your_env_name 文件可以在 Anaconda 安装目录 envs 文件下找到，也可在 Anaconda 软件中单击 Create 直接创建。

（4）使用激活（或切换不同 Python 版本）的虚拟环境。打开命令行输入 python --version 可以检查当前 Python 的版本。使用如下命令即可激活虚拟环境（即改变 Python 的版本）。Linux：source activate *your_env_name*（虚拟环境名称）。Windows：activate *your_env_name*（虚拟环境名称）。这时再使用 python --version 可以检查当前 Python 版本是否为想要的。

（5）在对虚拟环境中安装额外的包。

使用命令 conda install -n *your_env_name* [*package*] 即可安装 *package* 到 *your_env_name* 中。

（6）关闭虚拟环境（即从当前环境退出，返回使用 PATH 环境中的默认 Python 版本）。使用如下命令即可：Linux：source deactivate/Windows：deactivate。

（7）删除虚拟环境。使用命令 conda remove -n *your_env_name*（虚拟环境名称）-all，即可删除。

（8）删除环境中的某个包。使用命令 conda remove --name *your_env_name package_name* 即可。

1.3 TensorFlow 使用简介

TensorFlow 是 Google 公司开源的第二代用于数字计算的软件库。起初它是为了研究机器学习和深度神经网络而开发的，但后来发现这个系统足够通用，能够支持更加广泛的应用，就将其开源贡献了出来，其名字的由来也有源可溯——Tensor 意为张量，Flow 意为流，计算方法为从数据流图的一端流动到另一端[2]。

概括地说，TensorFlow 可以理解为一个深度学习框架，里面有完整的数据流向与处理机制，同时还封装了大量高效可用的算法及神经网络搭建方面的函数，可以在此基础之上进行深度学习的开发与研究。TensorFlow 是当今深度学习领域中影响力最大的框架之一。据相关统计，在 GitHub 上，TensorFlow 的受欢迎深度目前排名第一，以 3 倍左右的安装数量遥遥领先于第二名。

TensorFlow 是用 C++ 语言开发的，支持 C、Java、Python 等多种语言的调用，目前主流的方式通常会使用 Python 语言来驱动应用。这一特点也是其能够广受欢迎的原因。利用 C++ 语言开发可以保证其运行效率，Python 作为上层语言，可以为研究人员节省大量的开发时间。

1.3.1 TensorFlow 的特点

相对于其他框架，TensorFlow 有如下特点。

1. 灵活

TensorFlow 允许用户封装自己的"上层库"，所以即使在不需要使用底层语言的情况下，也能开发出新的复杂层类型。其采用的计算方法可以表示为一个数据流图，所以其特点是基于图运算，通过图上的各个节点变量控制训练中各个环节的变量，特别是需要对底层进行操作时，TensorFlow 比其他框架更容易。TensorFlow 的可移植性好，可以在 CPU 和 GPU 上运行，也可以在服务器、移动端、云端服务器等各种环境中运行。

2. 便捷和通用

TensorFlow 是现在的主流框架，使用 TensorFlow 生成的模型便捷、通用，在绝大多数的情况下都能满足使用者的需求。而且 TensorFlow 可以在多种系统下进行开发，例如 Linux、macOS、Windows 等。其编译好的模型几乎适用于当今所有的平台，几乎可以说是"即学即用"，模型易于学习和理解，使得用户应用起来更简单。

3. 框架成熟

众所周知，在 Google 公司内部大量的产品几乎都用到了 TensorFlow，如搜索排序、语音识别和自然语言处理等，TensorFlow 的各种优点使其得到广泛应用，由此可见该框架的成熟度较高。

4. 强悍的运算性能

TensorFlow 在大型计算机集群的并行处理中，运算性能仅略低于 CNTK，但在个人机器使用场景下会根据机器的配置自动选择 CPU 或 GPU 来运算，这方面做得更加友好和智能化。

1.3.2 CPU 版 TensorFlow 环境搭建与调用

TensorFlow 框架支持多种常见的操作系统，如 Windows 10、Ubuntu 18.04、macOS 等，同时也支持运行在 NVIDIA 显卡上的 GPU 版本和仅适用 CPU 完成计算的 CPU 版本。下面以常见的 Windows 10 系统、NVIDIA GPU、Python 语言环境为例，介绍如何安装 TensorFlow 框架及其他开发软件等。

本节介绍 CPU 版 TensorFlow 环境搭建与调用，1.3.3 节介绍 GPU 版 TensorFlow 环境搭建与调用。

下面介绍通过 Anaconda 方式安装 CPU 版 TensorFlow 的过程。这里需要 Python 64-bit，以及 Anaconda。

（1）创建虚拟环境。打开 Anaconda，单击 Create，创建 TensorFlow 环境，Python 版本选择 3.7，如图 1-9 所示。

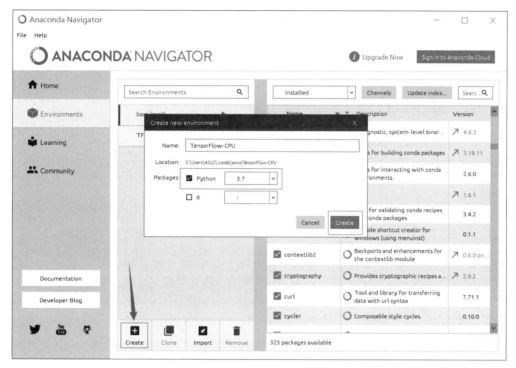

图 1-9　创建 TensorFlow 的 conda 环境

（2）安装 TensorFlow。在 cmd 命令行输入 activate TensorFlow-CPU 进入虚拟环境。现在来安装 TensorFlow CPU 最新版本。

```
#使用清华源安装 TensorFlow CPU 版本
pip install -U tensorflow-cpu==2.3.0 -i https://pypi.tuna.tsinghua.edu.cn/simple
```

上述命令自动下载 TensorFlow CPU 版本并安装，如图 1-10 所示，目前是 TensorFlow 2.3.0 正式版。-U 参数指定如果已安装此包，则执行升级命令。

（3）测试 TensorFlow 是否安装成功，先在窗口输入命令 python，执行后如图 1-11 所示。

（4）进入 TensorFlow 环境中的 Python，再输入命令：import tensorflow as tf，如果没有发生错误则表示已经安装成功。

1.3.3　GPU 版 TensorFlow 环境搭建与调用

一般来说，开发环境安装分为 4 个步骤：安装 Python 解释器 Anaconda，安装 CUDA 加速库，安装 TensorFlow 框架，安装常用编辑器。

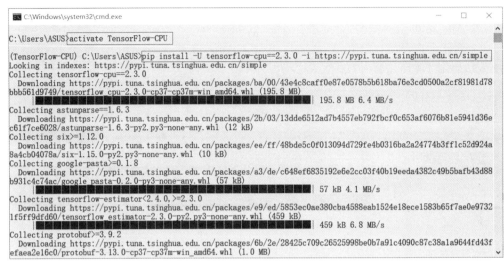

图 1-10　安装 TensorFlow

图 1-11　运行结果

TensorFlow 既可以支持 CPU，也可以支持 CPU＋GPU。前者的环境需求简单，后者需要额外的支持。如果要安装 GPU 版本，需要有 NVIDIA 显卡（俗称"N 卡"），以下为额外环境。

（1）有支持 CUDA 计算能力 3.0 或更高版本的 NVIDIA GPU 卡。

（2）下载安装 CUDA Toolkit，并确保其路径添加到 PATH 环境变量里。

（3）下载安装 cuDNN，并确保其路径添加到 PATH 环境变量里。

（4）CUDA 相关的 NVIDIA 驱动。

上述环境有一定的配对关系，存在不同的版本搭配，不可随意组合（在此以安装 TensorFlow 2.3 为例，更新的版本搭配可到 TensorFlow 官网查询）。

目前的深度学习框架大都基于 NVIDIA 的 GPU 显卡进行加速运算，因此需要安装 NVIDIA 提供的 GPU 加速库 CUDA 程序[3]。在安装 CUDA 之前，请确认本地计算机具有支持 CUDA 程序的 NVIDIA 显卡设备，如果计算机没有 NVIDIA 显卡，如部分计算机显卡为 AMD 以及部分 MacBook 笔记本电脑，则无法安装 CUDA 程序，因此可以跳过这一步，直接进入 TensorFlow 安装。

CUDA 的安装分为 CUDA 软件的安装、cuDNN 深度神经网络加速库的安装和环境变量配置 3 个步骤，安装稍微烦琐，请读者在操作时思考每个步骤的原因，避免死记硬背流程。

1. CUDA 软件安装

打开 CUDA 程序下载官网 https://developer.nvidia.com/cuda-toolkit-archive。这里使用 CUDA 10.1 版本，依次选择 Windows 平台、x86_64 架构、Windows 10 系统、exe(local)本地安装包，再单击 Download 即可下载 CUDA 安装软件。下载完成后，打开安装软件。如图 1-12 所示，选择"自定义"选项。

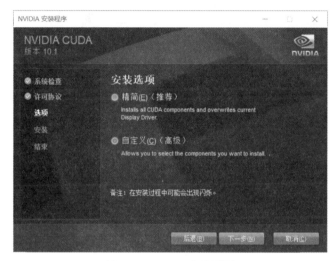

图 1-12　CUDA 安装界面

单击"下一步"按钮进入安装程序选择列表，在这里选择需要安装和取消不需要安装的程序。在 CUDA 节点下，取消 Visual Studio Integration 一项，如图 1-13 所示。在 Driver components 节点下，比对目前计算机已经安装的显卡驱动 Display Driver 的版本号"当前版本"和 CUDA 自带的显卡驱动版本号"新版本"，如果"当前版本"大于"新版本"，则需要取消 Display Driver 的勾，如图 1-14 所示；如果小于或等于，则默认勾选即可。设置完成后即可正常安装完成。

图 1-13　CUDA 安装界面（一）

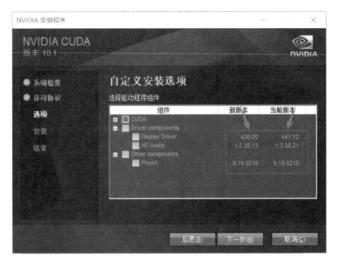

图 1-14　CUDA 安装界面（二）

安装完成后，需要测试 CUDA 软件是否安装成功。打开 cmd 命令行，输入 nvcc -V 并运行，即可打印当前 CUDA 的版本信息，如图 1-15 所示，如果命令无法识别，则说明安装失败。同时也可从 CUDA 的安装路径 C:\Program Files\NVIDIA GPU ComputingToolkit\CUDA\v10.0\bin 下找到 nvcc.exe 程序。

图 1-15　CUDA 安装结果测试

2. cuDNN 神经网络加速库安装

CUDA 并不是针对神经网络设计的 GPU 加速库，而是面向各种需要并行计算的应用设计。如果希望针对神经网络应用加速，需要额外安装 cuDNN 库。需要注意的是，cuDNN 库并不是运行程序，需要下载解压 cuDNN 文件，并配置 Path 环境变量。

打开网址 https://developer.nvidia.com/cudnn，选择 Download cuDNN，在此选择与 CUDA 10.1 匹配的 cuDNN 版本，并单击 cuDNN Library for Windows 10 链接即可下载 cuDNN 文件。需要注意的是，cuDNN 本身具有一个版本号，同时它还需要和 CUDA 的版本号对应上，不能下错不匹配 CUDA 版本号的 cuDNN 文件，如图 1-16 所示。

cuDNN 文件下载完成后，解压并进入文件夹，把目录 bin 中的 cudnn64_7.dll 复制到 C:\Program Files\NVIDIA GPU Computing Toolkit\CUDA\v10.1\bin 文件夹中。此处可能

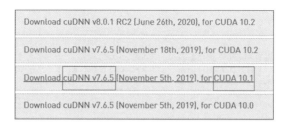

图 1-16 cuDNN 版本选择界面

会弹出需要管理员权限的对话框，选择继续即可粘贴，如图 1-17 所示。

图 1-17 cuDNN 文件的安装

CUDA 安装完成后，环境变量中应该包含 C:\Program Files\NVIDIA GPU Computing Toolkit\CUDA\v10.1\bin 和 C:\Program Files\NVIDIA GPU Computing Toolkit\CUDA\v10.1\libnvvp 两项，具体的路径可能依据实际路径略有出入，如图 1-18 所示，确认无误后依次单击"确定"按钮，关闭所有对话框。

3. TensorFlow 安装

创建虚拟环境，步骤同 CPU 安装步骤。使用 activate *your_envs* 命令进入创建的虚拟环境，然后输入以下命令：

```
#使用清华源安装 TensorFlow GPU 版本
pip install -U tensorflow-gpu==2.3.0 -i https://pypi.tuna.tsinghua.edu.cn/simple
```

上述命令自动下载 TensorFlow GPU 版本并安装，目前是 TensorFlow 2.3.0 正式版，-U 参数指定如果已安装此包，则执行升级命令。

现在来测试 GPU 版本的 TensorFlow 是否安装成功。在 cmd 命令行输入 python 进入 ipython 交互式终端，输入 import tensorflow as tf 命令，如果没有错误产生，继续输入 tf.

图 1-18　CUDA 相关的环境变量

test.is_gpu_available()测试 GPU 是否可用，此命令会打印出一系列以 I 开头的信息（Information），其中包含了可用的 GPU 显卡设备信息，最后会返回 True 或者 False，代表了 GPU 设备是否可用，如图 1-19 所示。如果为 True，则 TensorFlow GPU 版本安装成功；如果为 False，则安装失败，需要再次检测 CUDA、cuDNN、环境变量等步骤，或者复制错误提示再从搜索引擎中寻求帮助。

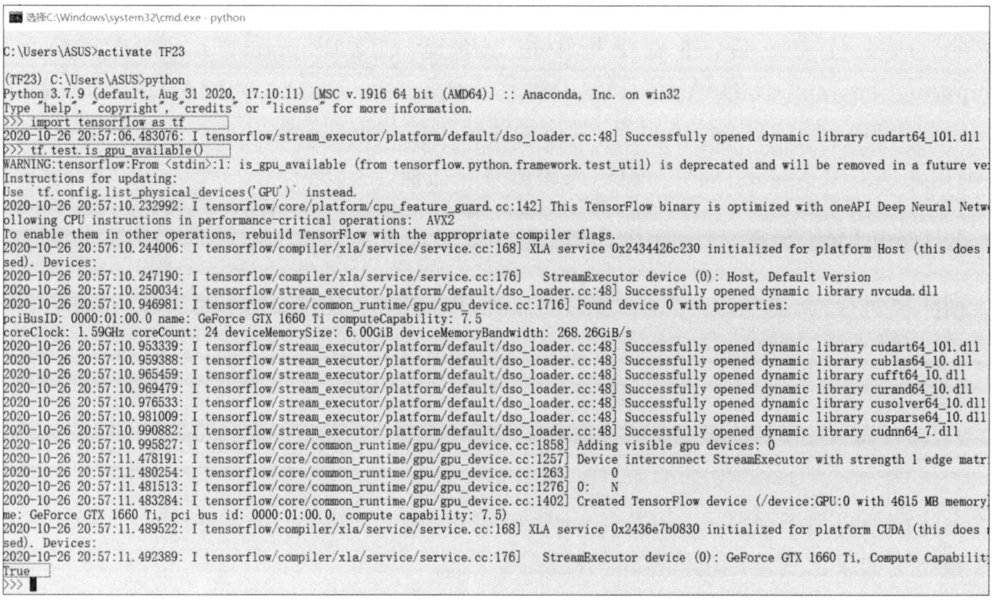

图 1-19　TensorFlow GPU 版本安装结果测试

1.4 Jupyter Notebook 使用简介

Jupyter Notebook 是一种 Web 应用,它能让用户将说明文本、数学方程、代码和可视化内容全部组合到一个易于共享的文档中,非常方便研究和教学[4]。在原始的 Python Shell 与 IPython 中,可视化在单独的窗口中进行,而文本资料以及各种函数和类脚本包含在独立的文档中。但是,Notebook 能将这一切集中到一处,让用户一目了然。

这里简单介绍如何使用 Jupyter Notebook 完成 TensorFlow 实战案例,更详细的使用方法可参考官方的说明。

1.4.1 安装 Jupyter Notebook

在创建好虚拟环境之后,打开终端,并激活之前创建的 Anaconda 环境。接下来要安装 Jupyter Notebook,在终端输入如下命令安装:

```
conda install jupyter notebook
```

或

```
pip install jupyter
```

安装完成后,若有任何 Jupyter Notebook 命令的疑问,可以考虑查看官方帮助文档,命令如下:

```
jupyter notebook -help
```

或

```
jupyter notebook -h
```

1.4.2 运行 Jupyter Notebook

在终端输入以下命令:

```
jupyter notebook
```

执行命令之后,在终端中将会显示一系列 Notebook 的服务器信息,同时浏览器将会自动启动 Jupyter Notebook。启动过程中终端显示内容如图 1-20 所示。

之后在 Jupyter Notebook 的所有操作,请保持终端不要关闭,因为一旦关闭终端,就会断开与本地服务器的链接,将无法在 Jupyter Notebook 中进行其他操作。

当执行完命令后,浏览器会进入 Jupyter Notebook 的主界面,单击 New→Python 3 即可创建新文件,如图 1-21 所示。Jupyter 编辑界面如图 1-22 所示。

注意:在 Jupyter Notebook 中编写代码所需的第三方库,都可在终端中通过 pip install 命令进行安装。

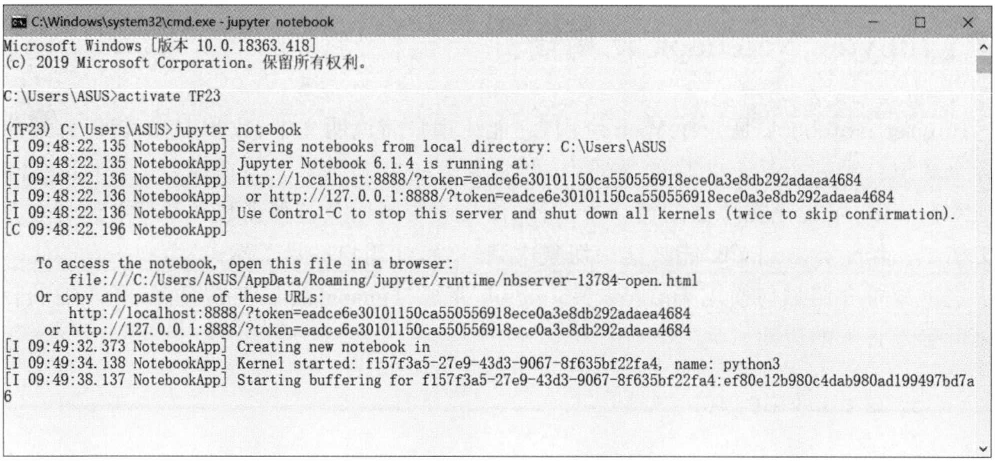

图 1-20 Jupyter Notebook 启动界面

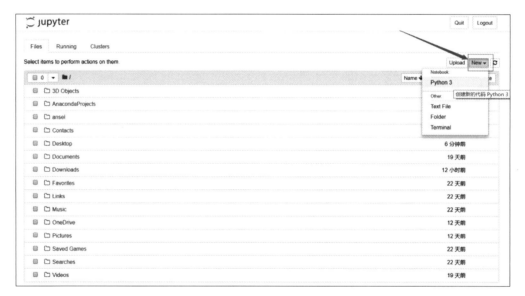

图 1-21 Jupyter 主界面

图 1-22 Jupyter 编辑界面

第 2 章

基本流程——数据处理与模型加载

本章将对深度学习基本流程进行介绍,包括数据的分类处理、回归方法的运用、过拟合与欠拟合的介绍以及模型的保存预加载。通过 4 个 TensorFlow 的实战案例,学会使用特征列队结构化数据进行分类,了解回归方法、过拟合与欠拟合,通过训练保存预训练模型索引。

2.1 使用特征列队结构化数据进行分类:预测心脏病

本次实战以 CSV 中的表格数据为例演示如何对结构化数据进行分类。

2.1.1 背景原理

tensorflow.keras 是用于构建和训练深度学习模型的 TensorFlow 高阶 API[5]。利用此 API,可实现快速原型设计、前沿研究和产品实现。它主要具有 3 大优势。

(1)方便用户使用:Keras 具有针对常见用例做出优化的简单而一致的界面。它可针对用户错误提供切实可行的清晰反馈。

(2)模块化和可组合:将可配置的构造块组合在一起就可以构建 Keras 模型,并且几乎不受限制。

(3)易于扩展:可以编写自定义构造块,表达新的研究创意;并且可以创建新层、指标、损失函数并开发先进的模型。

下面使用 Keras 定义、构建、训练和评估模型。

1. 对结构化数据进行分类

使用 Keras 来定义模型,将特征列(Feature Columns)作为从 CSV 中的列(Columns)映射到用于训练模型的特征(Features)桥梁。本次实战包括了以下内容的完整过程。

(1)用 Pandas 导入 CSV 文件。
(2)用 tf.data 建立输入流水线(Pipeline),用于对行进行分批(Batch)和随机排序(Shuffle)。
(3)用特征列将 CSV 中的列映射到用于训练模型的特征。
(4)用 Keras 构建、训练并评估模型。

2. 数据集

在此将使用一个小型数据集,该数据集由克利夫兰心脏病诊所基金会(Cleveland Clinic

Foundation for Heart Disease)提供[6]。CSV 中有几百行数据。每行描述了一个病人(Patient),每列描述了一个属性(Attribute)。将使用这些信息来预测一位病人是否患有心脏病,这是在该数据集上的二分类任务。表 2-1 是该数据集字段的描述。

表 2-1 数据集字段

列	描 述	特 征 类 型	数 据 类 型
Age	年龄以年为单位	数值	integer
Sex	性别(1 = 男;0 = 女)	类型	integer
CP	胸痛类型(0,1,2,3,4)	类型	integer
Trestbpd	静息血压(入院时,以 mmHg 计)	数值	integer
Chol	血清胆固醇(mg/dl)	数值	integer
FBS	空腹血糖($>$ 120 mg/dl)(1 = true;0 = false)	类型	integer
RestECG	静息心电图结果(0,1,2)	类型	integer
Thalach	达到的最大心率	数值	integer
Exang	运动诱发心绞痛(1 = 是;0 = 否)	类型	integer
Oldpeak	与休息时相比由运动引起的 ST 节段下降	数值	integer
Slope	在运动高峰 ST 段的斜率	数值	float
CA	荧光透视法染色的大血管动脉(0~3)的数量	数值	integer
Thal	3 表示正常;6 表示固定缺陷;7 表示可逆缺陷	类型	string
Target	心脏病诊断(1 = true;0 = false)	分类标注	integer

2.1.2 安装操作

本次实战的代码为 feature_columns.ipynb。首先进入 Anaconda Prompt 命令行,在本章根目录下输入命令 jupyter notebook,然后在浏览器中打开 feature_columns.ipynb,最后分步运行代码,即可得到程序运行结果。

1. 安装并导入 TensorFlow 和第三方库

```
import numpy as np
import pandas as pd
import tensorflow as tf
from tensorflow import feature_column
from tensorflow.keras import layers
from sklearn.model_selection import train_test_split
```

2. 使用 Pandas 创建一个 dataframe

Pandas 是一个 Python 库,它包含许多有用的实用程序,用于加载和处理结构化数据。在此 Pandas 从 URL 下载数据集,并将其加载到 dataframe 中。

```
URL = 'https://storage.googleapis.com/applied-dl/heart.csv'
dataframe = pd.read_csv(URL)
dataframe.head()
```

表 2-2 中显示了 5 个病人的各项属性信息，需要根据前 13 项属性判断该病人是不是心脏病患者，最后一列是该病人是否患心脏病的真实标记，1 表示是心脏病患者，0 表示不是心脏病患者。

表 2-2　用户基本数据表

	Age	Sex	CP	Trestbpd	Chol	FBS	RestECG	Thalach	Exang	Oldpeak	Slope	CA	Thal	Target
0	63	1	1	145	233	1	2	150	0	2.3	3	0	fixed	0
1	67	1	4	160	286	0	2	108	1	1.5	2	3	normal	1
2	67	1	4	120	229	0	2	129	1	2.6	2	2	reversible	0
3	37	1	3	130	250	0	0	187	0	3.5	3	0	normal	0
4	41	0	2	130	204	0	2	172	0	1.4	1	0	normal	0

2.1.3　代码解析

1. 拆分 dataframe

下载的数据集是一个 CSV 文件。将其拆分为训练集、验证集和测试集。

```
train, test = train_test_split(dataframe, test_size=0.2)
train, val = train_test_split(train, test_size=0.2)
print(len(train), 'train examples')
print(len(val), 'validation examples')
print(len(test), 'test examples')
```

2. 用 tf.data 创建输入流水线

接下来，将使用 tf.data 包装 dataframe。这能将特征列作为一座桥梁，该桥梁将 Pandas dataframe 中的列映射到用于训练模型的特征。如果使用一个非常大的 CSV 文件（非常大以至于它不能放入内存），将使用 tf.data 直接从磁盘读取它。本案例不涉及这一点。

```
#一种从 Pandas dataframe 创建 tf.data 数据集的实用程序方法(Utility Method)
def df_to_dataset(dataframe, shuffle=True, batch_size=32):
    dataframe = dataframe.copy()
    labels = dataframe.pop('target')
    ds = tf.data.Dataset.from_tensor_slices((dict(dataframe), labels))
    if shuffle:
        ds = ds.shuffle(buffer_size=len(dataframe))
    ds = ds.batch(batch_size)
    return ds
batch_size = 5                              #小批量大小用于演示
```

```
train_ds = df_to_dataset(train, batch_size=batch_size)
val_ds = df_to_dataset(val, shuffle=False, batch_size=batch_size)
test_ds = df_to_dataset(test, shuffle=False, batch_size=batch_size)
```

3. 理解输入流水线

现在已经创建了输入流水线,可以调用它来查看返回数据的格式。在此使用了一小批量大小来保持输出的可读性。

```
for feature_batch, label_batch in train_ds.take(1):
print('Every feature:', list(feature_batch.keys()))
print('A batch of ages:', feature_batch['age'])
print('A batch of targets:', label_batch )
Every feature: ['age', 'sex', 'cp', 'trestbps', 'chol', 'fbs', 'restecg', 'thalach
', 'exang','oldpeak', 'slope', 'ca', 'thal']
A batch of ages: tf.Tensor([58 44 44 50 54], shape=(5,), dtype=int64)
A batch of targets: tf.Tensor([0 1 0 0 1], shape=(5,), dtype=int64
```

可以看到数据集返回了一个字典,该字典从列名称(来自 dataframe)映射到 dataframe 中行的列值。

4. 演示几种特征列

TensorFlow 提供了多种特征列。本节中将创建几种特征列,并演示特征列如何转换 dataframe 中的列。

```
#将使用该批数据演示几种特征列
example_batch = next(iter(train_ds))[0]
#用于创建一个特征列
#并转换一批次数据的一个实用程序方法
def demo(feature_column):
    feature_layer = layers.DenseFeatures(feature_column)
    print(feature_layer(example_batch).numpy())
```

1)数值列

一个特征列的输出将成为模型的输入(使用上面定义的 demo 函数,将能准确地看到 dataframe 中每列的转换方式)。数值列(Numeric Column)是最简单的列类型。它用于表示实数特征。使用此列时,模型将从 dataframe 中接收未更改的列值。

```
age = feature_column.numeric_column("age")
demo(age)
```

在这个心脏病数据集中,dataframe 中的大多数列都是数值列。

2)分桶列

通常不希望将数字直接输入模型,而是根据数值范围将其值分成不同的类别。考虑代表一个人年龄的原始数据,可以用分桶列(Bucketized Column)将年龄分成几个分桶(Buckets),而不是将年龄表示成数值列。请注意下面的 one-hot 数值表示每行匹配的年龄范围。

```
age_buckets = feature_column.bucketized_column(age, boundaries=[18, 25, 30, 35,
40, 45, 50, 55, 60, 65])
demo(age_buckets)
```

3）分类列

在此数据集中，thal 用字符串表示（如 'fixed'、'normal'或 'reversible'）。无法直接将字符串提供给模型。相反，必须首先将它们映射到数值。分类词汇列（Categorical Vocabulary Columns）提供了一种用 one-hot 向量表示字符串的方法（就像在上面看到的年龄分桶一样）。词汇表可以用 categorical_column_with_vocabulary_list 作为 list 传递，或者用 categorical_column_with_vocabulary_file 从文件中加载。

```
thal = feature_column.categorical_column_with_vocabulary_list('thal', ['fixed',
'normal', 'reversible'])
thal_one_hot = feature_column.indicator_column(thal)
demo(thal_one_hot)
```

在更复杂的数据集中，许多列都是分类列（如 strings）。在处理分类数据时，特征列最有价值。尽管在该数据集中只有一列分类列，但将使用它来演示在处理其他数据集时，可以使用的几种重要的特征列。

4）嵌入列

假设不是只有几个可能的字符串，而是每个类别有数千（或更多）值。由于多种原因，随着类别数量的增加，使用 one-hot 编码训练神经网络变得不可行。可以使用嵌入列来克服此限制。嵌入列（Embedding Column）将数据表示为一个低维度密集向量，而非多维的 one-hot 向量，该低维度密集向量可以包含任何数，而不仅仅是 0 或 1。嵌入的大小（在下面的示例中为 8）是必须调整的参数。

当分类列具有许多可能的值时，最好使用嵌入列。在这里使用嵌入列用于演示目的，为了有一个完整的示例，以在将来可以修改用于其他数据集。

```
#注意到嵌入列的输入是之前创建的类别列
thal_embedding = feature_column.embedding_column(thal, dimension=8)
demo(thal_embedding)
```

5）经过哈希处理的特征列

表示具有大量数值的分类列的另一种方法，是使用 categorical_column_with_hash_bucket。该特征列计算输入的一个哈希值，然后选择一个 hash_bucket_size 分桶来编码字符串。使用此列时不需要提供词汇表，并且可以选择使 hash_buckets 的数量远远小于实际类别的数量以节省空间。

该技术的一个重要缺点是可能存在冲突，不同的字符串被映射到同一个范围。实际上，无论如何，经过哈希处理的特征列对某些数据集都有效。

```
thal_hashed = feature_column.categorical_column_with_hash_bucket('thal', hash_
bucket_size=1000)
demo(feature_column.indicator_column(thal_hashed))
```

6）组合的特征列

将多种特征组合到一个特征中，称为特征组合（Feature Crosses），它让模型能够为每种特征组合学习单独的权重。此处，将创建一个 age 和 thal 组合的新特征。请注意，crossed_column 不会构建所有可能组合的完整列表（可能非常大）。相反，它由 hashed_column 支持，因此可以选择表的大小。

```
crossed_feature = feature_column.crossed_column([age_buckets, thal], hash_bucket_size=1000)
demo(feature_column.indicator_column(crossed_feature))
```

前面已经了解了如何使用几种类型的特征列。现在将使用它们来训练模型。在此展示使用特征列所需的代码，任意地选择了几列来训练模型。

如果目标是建立一个准确的模型，请尝试使用自己的更大的数据集，并仔细考虑哪些特征最有意义，以及如何表示它们。

```
feature_columns = []
#数值列
for header in ['age', 'trestbps', 'chol', 'thalach', 'oldpeak', 'slope', 'ca']:
    feature_columns.append(feature_column.numeric_column(header))
#分桶列
age_buckets = feature_column.bucketized_column(age, boundaries=[18, 25, 30, 35, 40, 45, 50, 55, 60, 65])
feature_columns.append(age_buckets)
#分类列
thal = feature_column.categorical_column_with_vocabulary_list(
    'thal', ['fixed', 'normal', 'reversible'])
thal_one_hot = feature_column.indicator_column(thal)
feature_columns.append(thal_one_hot)
#嵌入列
thal_embedding = feature_column.embedding_column(thal, dimension=8)
feature_columns.append(thal_embedding)
#组合列
crossed_feature = feature_column.crossed_column([age_buckets, thal], hash_bucket_size=1000)
crossed_feature = feature_column.indicator_column(crossed_feature)
feature_columns.append(crossed_feature)
```

建立一个新的特征层：现在已经定义了特征列，将使用密集特征（Dense Features）层将特征列输入 Keras 模型中。

```
feature_layer = tf.keras.layers.DenseFeatures(feature_columns)
```

之前，使用一个小批量大小来演示特征列如何运转。下面将创建一个新的更大批量的输入流水线。

```
batch_size = 32
train_ds = df_to_dataset(train, batch_size=batch_size)
```

```
val_ds = df_to_dataset(val, shuffle=False, batch_size=batch_size)
test_ds = df_to_dataset(test, shuffle=False, batch_size=batch_size)
```

2.1.4　训练测试

下面创建编译和训练模型。

```
model = tf.keras.Sequential([
  feature_layer,
  layers.Dense(128, activation='relu'),
  layers.Dense(128, activation='relu'),
  layers.Dense(1, activation='sigmoid')
])
model.compile(optimizer='adam',
              loss='binary_crossentropy',
              metrics=['accuracy'],
              run_eagerly=True)
model.fit(train_ds,
          validation_data=val_ds,
          epochs=5)
loss, accuracy = model.evaluate(test_ds)
print("Accuracy", accuracy)
```

运行代码可看到训练过程和结果如下。

```
Epoch 1/5
7/7 [==============================] - 1s 95ms/step - loss: 1.7189 - accuracy: 0.6736 - val_loss: 0.0000e+00 - val_accuracy: 0.0000e+00
Epoch 2/5
7/7 [==============================] - 0s 45ms/step - loss: 0.8126 - accuracy: 0.6321 - val_loss: 0.7444 - val_accuracy: 0.7347
Epoch 3/5
7/7 [==============================] - 0s 49ms/step - loss: 1.2775 - accuracy: 0.7358 - val_loss: 1.1860 - val_accuracy: 0.7551
Epoch 4/5
7/7 [==============================] - 0s 49ms/step - loss: 0.7634 - accuracy: 0.7047 - val_loss: 0.8533 - val_accuracy: 0.5102
Epoch 5/5
7/7 [==============================] - 0s 48ms/step - loss: 0.6421 - accuracy: 0.7202 - val_loss: 0.7496 - val_accuracy: 0.7551
<tensorflow.python.keras.callbacks.History at 0x7fc68ac81c50>
loss, accuracy = model.evaluate(test_ds)print("Accuracy", accuracy)
2/2 [==============================] - 0s 26ms/step - loss: 0.8340 - accuracy: 0.6721
Accuracy 0.6721311
```

从训练过程和结果中可以看出，随着迭代次数的增加，模型的分类损失不断降低，准确率不断提高。当训练 5 个 epoch 后，模型的最终分类准确率达到 0.6721311，这个准确率并不是很高，因此通常使用更大、更复杂的数据集进行深度学习，将看到更好的结果。使用像这样的小数据集时，建议使用决策树或随机森林作为强有力的基准。本实战的目的不是训练一个准确的模型，而是演示处理结构化数据的机制，在将来使用自己的数据集时，就可以使用类似代码作为起点。

2.2 回归方法：预测燃油效率

本实战使用经典的 AutoMPG 数据集[7]，构建一个用来预测 20 世纪 70 年代末到 80 年代初汽车燃油效率的模型。

2.2.1 背景原理

在回归（Regression）问题中，目的是预测出如价格或概率这样连续值的输出。相对于分类（Classification）问题，分类的目的是从一系列的分类中选择出一个类别（例如，给出一张包含苹果或橘子的图像，识别出图像中是哪种水果）。

在本次实战中给该模型提供许多那个时期的汽车描述，这个描述包含气缸数、排量、马力以及质量等。

2.2.2 安装操作

本次实战的代码为 regression.ipynb。首先进入 Anaconda Prompt 命令行，在本章的根目录下输入命令 jupyter notebook，然后在浏览器中打开 regression.ipynb，最后分步运行代码，即可得到程序运行结果。安装并导入 TensorFlow 和第三方库。

```
import pathlib
import matplotlib.pyplot as plt
import pandas as pd
import seaborn as sns
import tensorflow as tf
from tensorflow import keras
from tensorflow.keras import layers
print(tf.__version__)
```

2.2.3 代码解析

1. AutoMPG 数据集

该数据集可以从 UCI 机器学习库中获取。

获取数据：首先下载数据集（见表 2-3）。数据集字段见表 2-4。

```
dataset_path = keras.utils.get_file("auto-mpg.data", "http://archive.ics.uci.edu/ml/machine-learning-databases/auto-mpg/auto-mpg.data")
dataset_path
```

再使用 pandas 导入数据集。

```
column_names = ['MPG','Cylinders','Displacement','Horsepower','Weight',
                'Acceleration', 'Model Year', 'Origin']
raw_dataset = pd.read_csv(dataset_path, names=column_names,
                    na_values = "?", comment='\t',
                    sep=" ", skipinitialspace=True)
dataset = raw_dataset.copy()
dataset.tail()
```

表 2-3　数据表项

	MPG	Cylinders	Displacement	Horsepower	Weight	Acceleration	Model Year	Origin
393	27.0	4	140.0	86.0	2790.0	15.6	82	1
394	44.0	4	97.0	52.0	2130.0	24.6	82	2
395	32.0	4	135.0	84.0	2295.0	11.6	82	1
396	28.0	4	120.0	79.0	2625.0	18.6	82	1
397	31.0	4	119.0	82.0	2720.0	19.4	82	1

表 2-4　数据集字段

列	描　　述	列	描　　述
MPG	miles per gallon, 每加仑行驶的英里数	Cylinders	气缸数
Displacement	发动机排量, 单位为升	Horsepower	马力
Weight	质量	Acceleration	加速度
Model Year	制造年份	Origin	生产产地

数据清洗：数据集中包括一些未知值。

```
dataset.isna().sum()
```

为了保证这个初始示例的简单性，删除这些行。

```
dataset = dataset.dropna()
```

Origin 列实际上代表分类，而不仅仅是一个数字，所以把它转换为独热码（one-hot）（见表 2-5）。

```
origin = dataset.pop('Origin')
dataset['USA'] = (origin == 1) * 1.0
dataset['Europe'] = (origin == 2) * 1.0
dataset['Japan'] = (origin == 3) * 1.0
dataset.tail()
```

表 2-5 转换为独热码的数据表项

	MPG	Cylinders	Displacement	Horsepower	Weight	Acceleration	Model Year	USA	Europe	Japan
393	27.0	4	140.0	86.0	2790.0	15.6	82	1.0	0.0	0.0
394	44.0	4	97.0	52.0	2130.0	24.6	82	0.0	1.0	0.0
395	32.0	4	135.0	84.0	2295.0	11.6	82	1.0	0.0	0.0
396	28.0	4	120.0	79.0	2625.0	18.6	82	1.0	0.0	0.0
397	31.0	4	119.0	82.0	2720.0	19.4	82	1.0	0.0	0.0

拆分训练数据集和测试数据集：现在需要将数据集拆分为一个训练数据集和一个测试数据集。最后将使用测试数据集对模型进行评估。

```
train_dataset = dataset.sample(frac=0.8,random_state=0)
test_dataset = dataset.drop(train_dataset.index)
```

数据检查：快速查看训练集中几对列的联合分布（见图 2-1）。

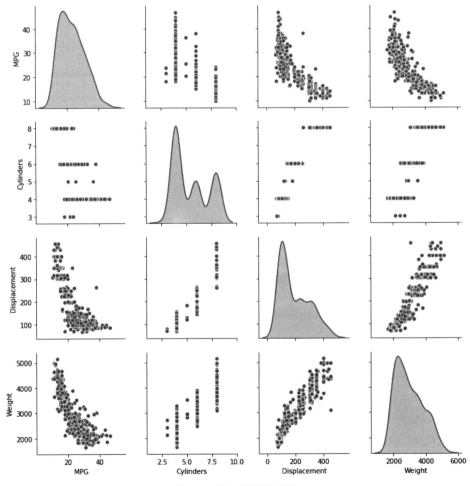

图 2-1 总体数据统计图

```
sns.pairplot(train_dataset[["MPG", "Cylinders", "Displacement", "Weight"]],
diag_kind="kde")
```

也可以查看总体的数据统计见表2-6。

```
train_stats = train_dataset.describe()
train_stats.pop("MPG")
train_stats = train_stats.transpose()
train_stats
```

表2-6　总体数据统计表

	count	mean	std	min	25%	50%	75%	max
Cylinders	314.0	5.477707	1.699788	3.0	4.00	4.0	8.00	8.0
Displacement	314.0	195.318471	104.331589	68.0	105.50	151.0	265.75	455.0
Horsepower	314.0	104.869427	38.096214	46.0	76.25	94.5	128.00	225.0
Weight	314.0	2990.251592	843.898596	1649.0	2256.50	2822.5	3608.00	5140.0
Acceleration	314.0	15.559236	2.789230	8.0	13.80	15.5	17.20	24.8
Model Year	314.0	75.898089	3.675642	70.0	73.00	76.0	79.00	82.0
USA	314.0	0.624204	0.485101	0.0	0.00	1.0	1.00	1.0
Europe	314.0	0.178344	0.383413	0.0	0.00	0.0	0.00	1.0
Japan	314.0	0.197452	0.398712	0.0	0.00	0.0	0.00	1.0

从标签中分离特征：将特征值从目标值或者标签中分离。这个标签是使用训练模型进行预测的值。

```
train_labels = train_dataset.pop('MPG')
test_labels = test_dataset.pop('MPG')
```

数据规范化：再次审视上面的train_stats部分，并注意每个特征的范围有什么不同。使用不同的尺度和范围对特征归一化是好的实践。尽管模型可能在没有特征归一化的情况下收敛，但它会使得模型训练更加复杂，并会造成生成的模型依赖输入所使用的单位选择。

尽管只从训练集中生成这些统计数据，但是这些统计信息也会用于归一化的测试数据集。需要将测试数据集放入到与已经训练过的模型相同的分布中。

```
def norm(x):
return (x - train_stats['mean']) / train_stats['std']
normed_train_data = norm(train_dataset)
normed_test_data = norm(test_dataset)
```

下面将会使用这个已经归一化的数据来训练模型。

用于归一化输入的数据统计（均值和标准差）需要反馈给模型从而应用于任何其他数据，以及之前所获得独热码。这些数据包含测试数据集以及生产环境中所使用的实时数据。

2. 模型

构建模型：下面构建自己的模型。这里将会使用一个顺序模型，其中包含两个紧密相连的隐藏层，以及返回单个、连续值的输出层。模型的构建步骤包含于一个名叫 build_model 的函数中，稍后将会创建第二个模型。第二个模型包括两个密集连接的隐藏层。

```
def build_model():
  model = keras.Sequential([
    layers.Dense(64, activation='relu', input_shape=[len(train_dataset.keys())]),
    layers.Dense(64, activation='relu'),
    layers.Dense(1)
  ])

  optimizer = tf.keras.optimizers.RMSprop(0.001)
  model.compile(loss='mse',
                optimizer=optimizer,
                metrics=['mae', 'mse'])
  return model
model = build_model()
```

检查模型：使用 .summary 方法来打印该模型的简单描述。

```
model.summary()
```

打印文本如下。

```
Model: "sequential"
_____
Layer (type)                 Output Shape              Param #
=================================================================
dense (Dense)                (None, 64)                640
_____
dense_1 (Dense)              (None, 64)                4160
_____
dense_2 (Dense)              (None, 1)                 65
=================================================================
Total params: 4,865
Trainable params: 4,865
Non-trainable params: 0
_____
```

现在试用一下这个模型。从训练数据中批量获取 10 条例子并对这些例子调用 model.predict。

```
example_batch = normed_train_data[:10]
example_result = model.predict(example_batch)
example_result
```

产生结果如下。

```
array([[ 0.01557792],
       [-0.09677997],
       [ 0.09817091],
       [-0.17441332],
       [-0.2521149 ],
       [-0.12406033],
       [-0.23704767],
       [-0.1016776 ],
       [-0.1110839 ],
       [-0.23353352]], dtype=float32)
```

产生了预期的形状和类型的结果。

2.2.4 训练测试

对模型进行 1000 个周期的训练，并在 history 对象中记录训练和验证的准确性。

```
#通过为每个完成的时期打印一个点来显示训练进度
class PrintDot(keras.callbacks.Callback):
  def on_epoch_end(self, epoch, logs):
    if epoch % 100 == 0: print('')
    print('.', end='')
EPOCHS = 1000
history = model.fit(
  normed_train_data, train_labels,
  epochs=EPOCHS, validation_split = 0.2, verbose=0,
  callbacks=[PrintDot()])
```

使用 history 对象中存储的统计信息可视化模型的训练进度。history 对象中存储的最后 5 条信息如表 2-7 所示。

```
hist = pd.DataFrame(history.history)
hist['epoch'] = history.epoch
hist.tail()
```

表 2-7 数据表项

序号	loss	mae	mse	val_loss	val_mae	val_mse	epoch
995	2.570732	1.051618	2.570732	10.587498	2.456362	10.587498	995
996	2.660562	1.022598	2.660562	10.711611	2.428715	10.711611	996
997	3.080793	1.141696	3.080793	10.469919	2.439384	10.469919	997
998	2.729193	1.066712	2.729193	10.671435	2.500411	10.671435	998
999	2.847594	1.041974	2.847594	10.892761	2.450137	10.892761	999

```
def plot_history(history):
```

```
    hist = pd.DataFrame(history.history)
    hist['epoch'] = history.epoch
    plt.figure()
    plt.xlabel('Epoch')
    plt.ylabel('Mean Abs Error [MPG]')
    plt.plot(hist['epoch'], hist['mae'],
             label='Train Error')
    plt.plot(hist['epoch'], hist['val_mae'],
             label = 'Val Error')
    plt.ylim([0,5])
    plt.legend()
    plt.figure()
    plt.xlabel('Epoch')
    plt.ylabel('Mean Square Error [$MPG^2$]')
    plt.plot(hist['epoch'], hist['mse'],
             label='Train Error')
    plt.plot(hist['epoch'], hist['val_mse'],
             label = 'Val Error')
    plt.ylim([0,20])
    plt.legend()
    plt.show()
plot_history(history)
```

图 2-2 和图 2-3 显示在约 100 个周期之后误差非但没有改进，反而出现恶化。下面更新 model.fit 调用，当验证值没有提高就自动停止训练。使用一个 EarlyStopping callback 来测试每个周期的训练条件。如果经过一定数量的周期后没有改进，则自动停止训练。

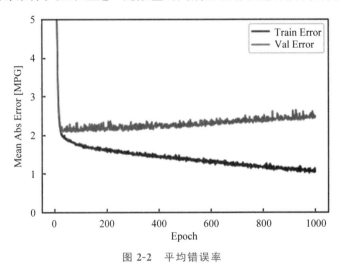

图 2-2　平均错误率

可以从这里学习到更多的回调。

```
model = build_model()
#patience 值用来检查改进 epochs 的数量
```

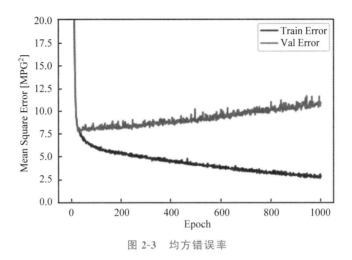

图 2-3　均方错误率

```
early_stop = keras.callbacks.EarlyStopping(monitor='val_loss', patience=10)

history = model.fit(normed_train_data, train_labels, epochs=EPOCHS,
                    validation_split = 0.2, verbose=0, callbacks=[early_stop,
PrintDot()])
plot_history(history)
```

平均错误率如图 2-4 所示。

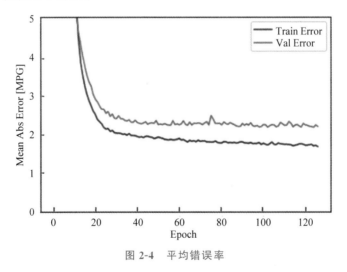

图 2-4　平均错误率

如图 2-5 所示，验证集中的平均的误差通常为±2MPG 左右。
下面看看通过使用测试集来泛化模型的效果如何，在训练模型时没有使用测试集。

```
loss, mae, mse = model.evaluate(normed_test_data, test_labels, verbose=2)
print("Testing set Mean Abs Error: {:5.2f} MPG".format(mae))

3/3 - 0s - loss: 5.9941 - mae: 1.8809 - mse: 5.9941
Testing set Mean Abs Error:  1.88 MPG
```

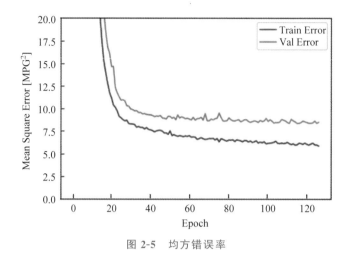

图 2-5 均方错误率

最后,使用测试集中的数据预测 MPG 值。

```
test_predictions = model.predict(normed_test_data).flatten()
plt.scatter(test_labels, test_predictions)
plt.xlabel('True Values [MPG]')
plt.ylabel('Predictions [MPG]')
plt.axis('equal')
plt.axis('square')
plt.xlim([0,plt.xlim()[1]])
plt.ylim([0,plt.ylim()[1]])
_ = plt.plot([-100, 100], [-100, 100])
```

图 2-6 显示出真实值与预测值分布相符,这看起来模型预测得相当好。

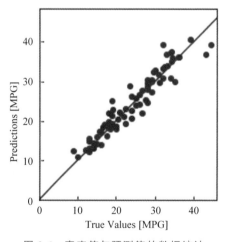

图 2-6 真实值与预测值的数据统计

下面来看看误差分布。

```
error = test_predictions - test_labels
```

```
plt.hist(error, bins = 25)
plt.xlabel("Prediction Error [MPG]")
_ = plt.ylabel("Count")
```

图 2-7 显示误差分布不是完全的高斯分布，但可以推断出，这是因为样本的数量很小所导致的。

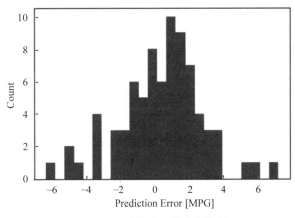

图 2-7　预测值失误的数据统计

2.3　过拟合与欠拟合：不同模型选择对结果的影响

在 2.2 节中实战示例中，可以看到模型在验证数据上的准确性在经过多个周期的训练后会达到峰值，然后会停滞不前或开始下降。

换句话说，模型将过拟合训练数据。学习如何处理过拟合是很重要的。尽管通常可以在训练集上达到高精度，但真正想要的是开发能够很好地推广到测试集（或之前未见的数据）的模型。

2.3.1　背景原理

过拟合的反面是欠拟合。当测试数据仍有改进空间时，就会发生欠拟合。发生这种情况的原因有很多：如果模型不够强大，模型过于规范化，或者仅仅是没有经过足够长时间的训练。这意味着网络尚未学习训练数据中的相关模式。

但是，如果训练时间过长，则模型将开始过拟合并从训练数据中学习无法推广到测试数据的模式。需要保持平衡。了解如何训练适当的周期是一项有用的技能。为了防止过拟合，最好的解决方案是使用更完整的训练数据。数据集应该涵盖模型预期处理的全部输入。额外的数据可能只有在涵盖新的和有趣的案例时才有用。

经过更完整数据训练的模型自然会更好地推广。如果做不到这一点，只好使用正则化之类的技术。这些因素限制了模型可以存储的信息的数量和类型。如果网络只能存储少量模式，那么优化过程将迫使它专注于最突出的模式，这些模式更有可能得到很好的推广。

2.3.2 安装操作

本次实战的代码为 overfit_and_underfit.ipynb。首先进入 Anaconda Prompt 命令行，在本章根目录下输入命令 jupyter notebook，然后在浏览器中打开 overfit_and_underfit.ipynb，最后分步运行代码，即可得到程序运行结果。

安装并导入 TensorFlow 和第三方库。

```
import tensorflow as t.
from tensorflow.keras import layers
from tensorflow.keras import regularizers
print(tf.__version__)
    !pip install -q git+https://github.com/tensorflow/docs
import tensorflow_docs as tfdocs
import tensorflow_docs.modeling
import tensorflow_docs.plots
    from  IPython import display
from matplotlib import pyplot as plt
import numpy as np
import pathlib
import shutil
import tempfile
```

2.3.3 代码解析

1. 希格斯数据集

该数据集包含 80 万个示例，每个示例有 28 个特性，以及一个二进制类标签[8]。

```
gz = tf.keras.utils.get_file('HIGGS.csv.gz', 'http://mlphysics.ics.uci.edu/data/higgs/HIGGS.csv.gz')
```

tf.data.experimental.CsvDataset 类可用于直接从 gzip 文件读取 csv 记录，而无需中间的解压缩步骤。

```
ds = tf.data.experimental.CsvDataset(gz,[float(),] * (FEATURES+1), compression_type="GZIP")
```

该 CSV 读取器类返回每个记录的标量列表。以下函数将标量列表重新打包为 (FEATURE_VECTOR, LABEL) 对。

```
def pack_row(* row):
label = row[0]
features = tf.stack(row[1:],1)
return features, label
```

TensorFlow 在处理大批量数据时效率最高。因此，与其单独重新打包每一行，不如创建一个新的数据集，该数据集采用 10000 个示例的批次，将 pack_row 函数应用于每个批次，然后将这些批次拆分回各个记录。

```
packed_ds = ds.batch(10000).map(pack_row).unbatch()
```

请看一下这个新的PACKED_DS的一些记录。图2-8显示了这些预测值的数据分布统计。这些功能没有完全规范化，但这对于本次实战来说已经足够了。

```
for features,label in packed_ds.batch(1000).take(1):
print(features[0])
plt.hist(features.numpy().flatten(), bins = 101)
```

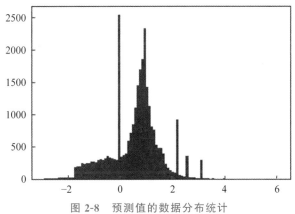

图2-8　预测值的数据分布统计

为了使本教程相对简短，只使用前1000个样本进行验证，然后使用接下来的10000个样本进行训练。

```
N_VALIDATION = int(1e3)
N_TRAIN = int(1e4)
BUFFER_SIZE = int(1e4)
BATCH_SIZE = 500
STEPS_PER_EPOCH = N_TRAIN//BATCH_SIZE
```

Dataset.skip和Dataset.take方法使此操作变得容易。

同时，使用Dataset.cache方法来确保加载器不需要在每个周期重新从文件中读取数据。

```
validate_ds = packed_ds.take(N_VALIDATION).cache()
train_ds = packed_ds.skip(N_VALIDATION).take(N_TRAIN).cache()
    train_ds
```

这些数据集返回单个示例。使用.batch方法可创建适当大小的批次进行训练。批处理之前，还需要使用.shuffle和.repeat方法将训练集打乱并进行一定的重复。

```
validate_ds = validate_ds.batch(BATCH_SIZE)
train_ds = train_ds.shuffle(BUFFER_SIZE).repeat().batch(BATCH_SIZE)
```

2. 演示过拟合

防止过拟合的最简单方法是从小模型开始：具有少量可学习参数（由层数和每层单元

数确定)的模型。在深度学习中,模型中可学习参数的数量通常被称为模型的"容量"。直观地说,参数越多的模型将具有越强的"记忆能力"。因此,将能够容易地学习训练样本与其目标之间的完美字典式映射,这种映射没有任何泛化能力,但当对以前未见的数据进行预测时,这将是无用的。

始终牢记这一点:深度学习模型往往擅长拟合训练数据,但真正的挑战是泛化,而不是拟合。另一方面,如果网络的记忆资源有限,它就不能那么容易地学习映射。为了将损失降到最低,它将不得不学习具有更强预测能力的压缩表示法。同时,如果模型太小,它将很难适应训练数据。"容量过大"和"容量不足"之间存在平衡。

不幸的是,没有特定的公式来确定模型的正确大小或体系结构(根据层的数量,或者每层的正确大小)。因此,将不得不使用一系列不同的体系结构进行实验。要找到合适的模型大小,最好从相对较少的层和参数开始,然后开始增加层的大小或添加新层,直到看到验证损失的回报递减。从简单模型开始,以密度作为基准,然后创建较大的版本,并对它们进行比较。

3. 训练程序

如果在训练过程中逐渐降低学习率,许多模型的训练效果会更好。使用 optimizers.schedules 随着时间的推移降低学习率。

```
lr_schedule = tf.keras.optimizers.schedules.InverseTimeDecay(
    0.001,
    decay_steps=STEPS_PER_EPOCH * 1000,
    decay_rate=1,
    staircase=False)
def get_optimizer():
    return tf.keras.optimizers.Adam(lr_schedule)
```

上面的代码设置了一个 schedules.InverseTimeDecay,以双曲线的方式将学习速率在 1000 个时代降低到基本速率的 1/2,在 2000 个时代降低 1/3,以此类推,如图 2-9 所示。

```
step = np.linspace(0,100000)
lr = lr_schedule(step)
plt.figure(figsize = (8,6))
plt.plot(step/STEPS_PER_EPOCH, lr)
plt.ylim([0,max(plt.ylim())])
plt.xlabel('Epoch')
_ = plt.ylabel('Learning Rate')
```

本书中的每个模型都将使用相同的训练配置。因此,从回调列表开始,以可重用的方式设置它们。

本书的训练会持续很短的时间。为了减少记录噪声,请使用 tfdocs.EpochDots,它打印一个对于每个周期,以及每 100 个时期的完整指标。

接下来包括 callbacks.EarlyStopping 以避免冗长和不必要的训练时间。请注意,此回调设置为监视 val_binary_crossentropy,而不是 val_loss。这种差异稍后将变得很重要。

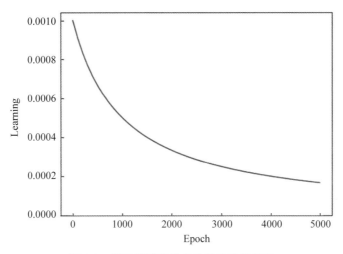

图 2-9　学习率随着迭代次数的改变情况

使用 callbacks.TensorBoard 生成用于训练的 TensorBoard 日志。

```
def get_callbacks(name):
  return [
    tfdocs.modeling.EpochDots(),
    tf.keras.callbacks.EarlyStopping(monitor='val_binary_crossentropy', patience=200),
    tf.keras.callbacks.TensorBoard(logdir/name),
  ]
```

同样，每个模型都将使用相同的 Model.Compile 和 Model.fit 设置：

```
def compile_and_fit(model, name, optimizer=None, max_epochs=10000):
  if optimizer is None:
    optimizer = get_optimizer()
  model.compile(optimizer=optimizer,
            loss=tf.keras.losses.BinaryCrossentropy(from_logits=True),
            metrics=[
              tf.keras.losses.BinaryCrossentropy(
                from_logits=True, name='binary_crossentropy'),
              'accuracy'])
  model.summary()
  history = model.fit(
    train_ds,
    steps_per_epoch = STEPS_PER_EPOCH,
    epochs=max_epochs,
    validation_data=validate_ds,
    callbacks=get_callbacks(name),
    verbose=0)
  return history
```

1)微型模型

从训练模型开始。

```
tiny_model = tf.keras.Sequential([
    layers.Dense(16, activation='elu', input_shape=(FEATURES,)),
    layers.Dense(1)
])
size_histories = {}
size_histories['Tiny'] = compile_and_fit(tiny_model, 'sizes/Tiny')
```

2)小模型

尝试 2 个各有 16 个单位的隐藏层。

```
small_model = tf.keras.Sequential([
    #`input_shape` is only required here so that `.summary` works.
    layers.Dense(16, activation='elu', input_shape=(FEATURES,)),
    layers.Dense(16, activation='elu'),
    layers.Dense(1)
    ])
    size_histories['Small'] = compile_and_fit(small_model, 'sizes/Small')
```

3)中型模型

现在尝试 3 个隐藏层,每个层有 64 个单位。

```
medium_model = tf.keras.Sequential([
    layers.Dense(64, activation='elu', input_shape=(FEATURES,)),
    layers.Dense(64, activation='elu'),
    layers.Dense(64, activation='elu'),
    layers.Dense(1)
])
```

并使用相同的数据训练模型。

```
size_histories['Medium']  = compile_and_fit(medium_model, "sizes/Medium")
```

4)大型模型

可以创建更大的模型作为练习,并查看它开始过拟合的速度有多快。接下来,在此基准中添加一个容量大得多的网络,其容量远远超过问题所需的容量。

```
large_model = tf.keras.Sequential([
    layers.Dense(512, activation='elu', input_shape=(FEATURES,)),
    layers.Dense(512, activation='elu'),
    layers.Dense(512, activation='elu'),
    layers.Dense(512, activation='elu'),
    layers.Dense(1)
])
```

并且,再次使用相同的数据训练模型。

```
size_histories['large'] = compile_and_fit(large_model, "sizes/large")
```

4. 绘制训练和验证损失

图 2-10 的实线表示训练损失，虚线表示测试得到的实际损失（请记住：实际损失越小表示模型越好）。

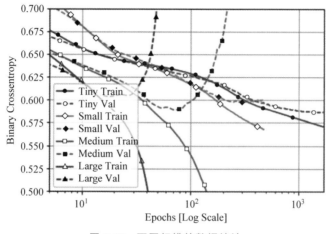

图 2-10　不同规模的数据统计

虽然构建一个更大的模型会给它带来更大的能力，但如果这种能力没有受到某种限制，它就会轻松地将其过拟合至训练集。在本例中，通常只有"微型"模型能够避免完全过拟合，而每个较大的模型都能更快地拟合数据。对于"大型"模型来说，这变得非常严重，以至于需要将绘图切换到对数尺度才能真正看到正在发生的事情。如果将验证指标与训练指标进行比较，这很明显。

有微小的差异是正常的。如果两个指标都朝着同一个方向发展，那么一切都正常；如果验证指标开始停滞不前，而训练指标继续改善，那么可能接近过拟合了；如果验证指标走错了方向，那么模型显然是过拟合了。

5. 绘制结果对比图

```
plotter.plot(size_histories)
a = plt.xscale('log')
plt.xlim([5, max(plt.xlim())])
plt.ylim([0.5, 0.7])
plt.xlabel("Epochs [Log Scale]")
```

2.3.4　训练测试

1. 添加权重正规化

正则化可能类似于 Razor 原理：给出对某事的两种解释，最有可能正确的解释是"最简单"的解释，也是做出假设最少的一种。这也适用于神经网络学习的模型：给定一些训练数据和网络结构,有多组权值(多个模型)可以解释数据,简单的模型与复杂的模型相比,不太可能过拟合。

在这种情况下,简单模型是参数值的分布具有较小熵的模型(或如上文所述,具有总共较少参数的模型)。因此,减轻过拟合的一种常见方法是通过强制网络的权值只取较小的值来限制网络的复杂性,这使得权值的分布更加"规则"。这被称为"权重正则化",并且这是通过向网络的损失函数添加与具有较大权重相关联的成本来实现的。此成本有两种类型。

(1) L1 正则化,其中添加的成本与权重系数的绝对值成正比(即与权重的 L1 范数成正比)。

(2) L2 正则化,其中增加的成本与权重系数值的平方成正比。在神经网络中,L2 正则化也称为权重衰减。不要让不同的名称产生混淆:权重衰减在数学上与 L2 正则化完全相同。

在 tf.keras 中,通过将权重正则化实例作为关键字参数传递给层来添加权重正则化。现在添加 L2 权重正则化。

```
l2_model = tf.keras.Sequential([
    layers.Dense(512, activation='elu',
                 kernel_regularizer=regularizers.l2(0.001),
                 input_shape=(FEATURES,)),
    layers.Dense(512, activation='elu',
                 kernel_regularizer=regularizers.l2(0.001)),
    layers.Dense(512, activation='elu',
                 kernel_regularizer=regularizers.l2(0.001)),
    layers.Dense(512, activation='elu',
                 kernel_regularizer=regularizers.l2(0.001)),
    layers.Dense(1)
])
regularizer_histories['l2'] = compile_and_fit(l2_model, "regularizers/l2")
```

l2(0.001)表示该层的权重矩阵中的每个系数都会使网络的总损耗增加 0.001 * weight_coefficient_value ** 2。这就是为什么要直接监控 binary_crossentropy 的原因。因为它没有包含到正则化成分中。因此,具有 L2 正则化惩罚的相同的"大"模型的性能要好得多。

绘制结果对比图。

```
plotter.plot(regularizer_histories)plt.ylim([0.5, 0.7])
```

可见,正则化的 L2 模型现在比微型模型更具竞争力。尽管具有相同数量的参数,但 L2 模型也比其基于的"大"模型更耐过拟合。

2. 添加丢失

丢失(Dropout)是神经网络中最有效和最常用的正则化技术之一。

丢失的直观解释是,因为网络中的单个节点不能依赖其他节点的输出,所以每个节点必须输出自己有用的功能。应用于图层的丢弃包括在训练过程中随机"丢失"(即设置为零)该图层的多个输出要素。假设在训练过程中,给定的图层通常会为给定的输入样本返回向量

[0.2、0.5、1.3、0.8、1.1];在应用丢失之后,此向量将有一些零个条目随机分布,例如[0,0.5,1.3,0,1.1]。丢失率是被清零的特征的一部分,通常设置为 0.2~0.5。在测试时,不会丢失任何单元,而是将图层的输出值按等于丢失率的比例缩小,以平衡比训练时活动的单元更多的事实。

在 tf.keras 中可以通过丢失层在网络中引入丢失,该层将立即应用于该层的输出。下面在网络中添加两个丢失层,看看它们在减少过拟合方面表现如何。

```
dropout_model = tf.keras.Sequential([
    layers.Dense(512, activation='elu', input_shape=(FEATURES,)),
    layers.Dropout(0.5),
    layers.Dense(512, activation='elu'),
    layers.Dropout(0.5),
    layers.Dense(512, activation='elu'),
    layers.Dropout(0.5),
    layers.Dense(512, activation='elu'),
    layers.Dropout(0.5),
    layers.Dense(1)
])
regularizer_histories['dropout'] = compile_and_fit(dropout_model,
"regularizers/dropout")plotter.plot(regularizer_histories)
plt.ylim([0.5, 0.7])
```

这两种正则化方法都改进了"大"模型的行为。但这仍然不能超越"微小"基准。接下来,将这两种方法一起尝试,看看是否效果更好。

```
combined_model = tf.keras.Sequential([
    layers.Dense(512, kernel_regularizer=regularizers.l2(0.0001),
                 activation='elu', input_shape=(FEATURES,)),
    layers.Dropout(0.5),
    layers.Dense(512, kernel_regularizer=regularizers.l2(0.0001),
                 activation='elu'),
    layers.Dropout(0.5),
    layers.Dense(512, kernel_regularizer=regularizers.l2(0.0001),
                 activation='elu'),
    layers.Dropout(0.5),
    layers.Dense(512, kernel_regularizer=regularizers.l2(0.0001),
                 activation='elu'),
    layers.Dropout(0.5),
    layers.Dense(1)
])
regularizer_histories['combined'] = compile_and_fit(combined_model,
"regularizers/combined")plotter.plot(regularizer_histories)
plt.ylim([0.5, 0.7])
```

经过实战可知以下是防止神经网络过拟合的最常用方法。

(1) 获取更多训练数据。

（2）减少网络容量。
（3）添加权重正规化。
（4）添加丢失。

本节未涵盖的两种重要方法是。

（1）数据扩充。
（2）批量标准化。

每种方法都可以单独提供帮助，但通常将它们组合起来会更加有效。

2.4　保存与加载预训练模型

本次实战将演示如何在训练过程中保存模型、如何保存整个模型、如何将模型保存为 HDF5 和 saved_model 格式，以及如何加载模型。

2.4.1　背景原理

模型可以在训练期间和训练完成后进行保存。这意味着模型可以从任意中断中恢复，并避免耗费比较长的时间在训练上。保存也意味着可以共享模型，其他人可以通过这些模型来重新创建工作。

本次实战中使用的数据集是 MNIST 数据集，这是一个入门级计算机视觉数据集，由 0~9 手写数字图像和数字标签组成，其中包含 60 000 个训练样本和 10 000 个测试样本，每个样本都是一张 28 像素×28 像素的灰度手写数字图像。实战目标是训练一个模型用于识别图像里面的数字，并得到较高的分类准确率。

2.4.2　安装操作

本次实战的代码为 save_and_load.ipynb。首先进入 Anaconda Prompt 命令行，在本章根目录下输入命令 jupyter notebook，然后在浏览器中打开 save_and_load.ipynb，最后分步运行代码，即可得到程序运行结果。

1. 安装并导入 TensorFlow 和第三方库

```
pip install -q pyyaml h5py            #以 HDF5 格式保存模型所必需
import os
import tensorflow as tf
from tensorflow import keras
```

2. 获取示例数据集

要演示如何保存和加载权重，将使用 MNIST 数据集。要加快运行速度，请使用前 1000 个示例。

```
(train_images, train_labels), (test_images, test_labels) = tf.keras.datasets.mnist.load_data()
train_labels = train_labels[:1000]
```

```
test_labels = test_labels[:1000]
train_images = train_images[:1000].reshape(-1, 28 * 28) / 255.0
test_images = test_images[:1000].reshape(-1, 28 * 28) / 255.0
```

2.4.3　代码解析

1. 定义模型

首先构建一个简单的序列（Sequential）模型。

```
#定义一个简单的序列模型
def create_model():
  model = tf.keras.models.Sequential([
    keras.layers.Dense(512, activation='relu', input_shape=(784,)),
    keras.layers.Dropout(0.2),
    keras.layers.Dense(10)
  ])
  model.compile(optimizer='adam',
                loss=tf.losses.SparseCategoricalCrossentropy(from_logits=True),
                metrics=['accuracy'])
  return model
#创建一个基本的模型实例
model = create_model()
#显示模型的结构
model.summary()
```

显示结果如下。

```
Model: "sequential"
_____
Layer (type)                 Output Shape              Param #
=================================================================
dense (Dense)                (None, 512)               401920
_____
dropout (Dropout)            (None, 512)               0
_____
dense_1 (Dense)              (None, 10)                5130
=================================================================
Total params: 407,050
Trainable params: 407,050
Non-trainable params: 0
_____
```

2. 在训练期间保存模型（以 checkpoints 形式保存）

可以使用训练好的模型而不需要从头开始重新训练，或在打断的地方开始训练，以防止训练过程没有保存。tf.keras.callbacks.ModelCheckpoint 允许在训练的过程中和结束时回调保存的模型。

Checkpoint 回调用法：创建一个只在训练期间保存权重的 tf.keras.callbacks.ModelCheckpoint 回调。

```
checkpoint_path = "training_1/cp.ckpt"
checkpoint_dir = os.path.dirname(checkpoint_path)
#创建一个保存模型权重的回调
cp_callback = tf.keras.callbacks.ModelCheckpoint(filepath=checkpoint_path,
                                                 save_weights_only=True,
                                                 verbose=1)
#使用新的回调训练模型
model.fit(train_images,
          train_labels,
          epochs=10,
          validation_data=(test_images,test_labels),
          callbacks=[cp_callback])    #通过回调训练
#这可能会生成与保存优化程序状态相关的警告
#这些警告(以及整个笔记本中的类似警告)是防止过时使用,可以忽略
```

这将创建一个 TensorFlow checkpoint 文件集合，这些文件在每个 epoch 结束时更新。

```
ls {checkpoint_dir}
```

3. 创建一个新的未经训练的模型

仅恢复模型的权重时，必须具有与原始模型相同网络结构的模型。由于模型具有相同的结构，可以共享权重，尽管它是模型的不同实例。现在重建一个新的未经训练的模型，并在测试集上进行评估。未经训练的模型将在机会水平（Chance Levels）上执行（准确度约为 10%）。

```
#创建一个基本模型实例
model = create_model()
#评估模型
loss, acc = model.evaluate(test_images,  test_labels, verbose=2)
print("Untrained model, accuracy: {:5.2f}%".format(100 * acc))
```

然后从 checkpoint 加载权重并重新评估。

```
#加载权重
model.load_weights(checkpoint_path)
#重新评估模型
loss,acc = model.evaluate(test_images,  test_labels, verbose=2)
print("Restored model, accuracy: {:5.2f}%".format(100 * acc))
```

checkpoint 回调选项：回调提供了几个选项，为 checkpoint 提供唯一名称并调整 checkpoint 频率。训练一个新模型，每 5 个周期保存一次唯一命名的 checkpoint。

```
#在文件名中包含 epoch (使用 `str.format`)
```

```
checkpoint_path = "training_2/cp-{epoch:04d}.ckpt"
checkpoint_dir = os.path.dirname(checkpoint_path)
#创建一个回调,每5个周期保存模型的权重
cp_callback = tf.keras.callbacks.ModelCheckpoint(
    filepath=checkpoint_path,
    verbose=1,
    save_weights_only=True,
    period=5)
#创建一个新的模型实例
model = create_model()
#使用 `checkpoint_path` 格式保存权重
model.save_weights(checkpoint_path.format(epoch=0))
#使用新的回调训练模型
model.fit(train_images,
          train_labels,
          epochs=50,
          callbacks=[cp_callback],
          validation_data=(test_images,test_labels),
          verbose=0)
```

现在查看生成的 checkpoint 并选择最新的 checkpoint。

```
! ls {checkpoint_dir}
    checkpoint                cp-0025.ckpt.index
    cp-0000.ckpt.data-00000-of-00001  cp-0030.ckpt.data-00000-of-00001
    cp-0000.ckpt.index         cp-0030.ckpt.index
    cp-0005.ckpt.data-00000-of-00001  cp-0035.ckpt.data-00000-of-00001
    cp-0005.ckpt.index         cp-0035.ckpt.index
    cp-0010.ckpt.data-00000-of-00001  cp-0040.ckpt.data-00000-of-00001
    cp-0010.ckpt.index         cp-0040.ckpt.index
    cp-0015.ckpt.data-00000-of-00001  cp-0045.ckpt.data-00000-of-00001
    cp-0015.ckpt.index         cp-0045.ckpt.index
    cp-0020.ckpt.data-00000-of-00001  cp-0050.ckpt.data-00000-of-00001
    cp-0020.ckpt.index         cp-0050.ckpt.index
    cp-0025.ckpt.data-00000-of-00001
latest = tf.train.latest_checkpoint(checkpoint_dir)
latest
    'training_2/cp-0050.ckpt'
```

注意:默认的 TensorFlow 格式仅保存最近的 5 个 checkpoint。

如果要进行测试,请重置模型并加载最新的 checkpoint。

```
#创建一个新的模型实例
model = create_model()
#加载以前保存的权重
model.load_weights(latest)
#重新评估模型
```

```
loss, acc = model.evaluate(test_images,  test_labels, verbose=2)
print("Restored model, accuracy: {:5.2f}%".format(100 * acc))
```

上述代码将权重存储到 checkpoint——格式化文件的集合中，这些文件仅包含二进制格式的训练权重。checkpoints 包含：一个或多个包含模型权重的分片；索引文件，指示哪些权重存储在哪个分片中。

如果只在一台机器上训练一个模型，将有一个带有后缀的碎片：.data-00000-of-00001。

4. 手动保存权重

下面将了解如何将权重加载到模型中。使用 model.save_weights 方法手动保存它们同样简单。默认情况下，tf.keras 和 save_weights 使用 TensorFlow checkpoints 格式保存权重，权重文件扩展名为 ckpt（若以 HDF5 格式保存并序列化模型，扩展名为 h5）。

```
#保存权重
model.save_weights('./checkpoints/my_checkpoint')
#创建模型实例
model = create_model()
#恢复权重
model.load_weights('./checkpoints/my_checkpoint')
#评估模型
loss,acc = model.evaluate(test_images,  test_labels, verbose=2)
print("Restored model, accuracy: {:5.2f}%".format(100 * acc))
32/32 - 0s - loss: 0.4836 - accuracy: 0.8750
Restored model, accuracy: 87.50%
```

5. 保存整个模型

模型和优化器可以保存到包含其状态（权重和变量）和模型参数的文件中。这可以导出模型，以便在不访问原始 Python 代码的情况下使用它，而且可以通过恢复优化器状态的方式，从中断的位置恢复训练。保存完整模型会非常有用——可以在 TensorFlow.js（HDF5，Saved Model）加载它们，然后在 Web 浏览器中训练和运行它们，或者使用 TensorFlowLite 将它们转换为在移动设备上运行（HDF5，Saved Model）。

将模型保存为 HDF5 文件：Keras 可以使用 HDF5 标准提供基本保存格式，可以将保存的模型视为单个二进制 blob。

```
#创建一个新的模型实例
model = create_model()
#训练模型
model.fit(train_images, train_labels, epochs=5)
#将整个模型保存为 HDF5 文件
model.save('my_model.h5')
```

现在，从该文件重新创建模型。

```
#重新创建完全相同的模型,包括其权重和优化程序
```

```
new_model = tf.keras.models.load_model('my_model.h5')
#显示网络结构
new_model.summary()
```

显示结果如下。

```
Model: "sequential_5

_____
Layer (type)              Output Shape         Param #
=================================================
dense_12 (Dense)          (None, 512)          401920

dropout_6 (Dropout)       (None, 512)          0

dense_13 (Dense)          (None, 10)           5130
=================================================
Total params: 407,050
Trainable params: 407,050
Non-trainable params: 0
```

检查其准确率(accuracy)。

```
loss, acc = new_model.evaluate(test_images, test_labels, verbose=2)
print('Restored model, accuracy: {:5.2f}%'.format(100 * acc))

32/32 - 0s - loss: 0.4639 - accuracy: 0.0840
Restored model, accuracy:  8.40%
```

这项技术可以保存一切：权重、模型配置(结构)、优化器配置。Keras 通过检查网络结构来保存模型。目前，它无法保存 TensorFlow 优化器(调用自 tf.train)。使用这些的时候，需要在加载后重新编译模型，否则将失去优化器的状态。

2.4.4 训练测试

训练好的模型通过 saved_model 保存。SavedModel 格式是序列化模型的另一种方法。以这种格式保存的模型，可以使用 tf.keras.models.load_model 还原，并且模型与 TensorFlow Serving 兼容。

```
#创建并训练一个新的模型实例
model = create_model()
model.fit(train_images, train_labels, epochs=5)
#将整个模型另存为 SavedModel
!mkdir -p saved_model
model.save('saved_model/my_model')
```

SavedModel 格式是一个包含 protobuf 二进制文件和 TensorFlow 检查点(Checkpoint)的目录。检查保存的模型目录。

```
#my_model 文件夹
```

```
!ls saved_model
#包含一个 assets 文件夹、saved_model.pb 和变量文件夹
!ls saved_model/my_model
my_model
assets   saved_model.pb  variables
```

从保存的模型重新加载新的 Keras 模型。

```
new_model = tf.keras.models.load_model('saved_model/my_model')
#检查其架构
new_model.summary()
```

显示结果如下。

```
Model: "sequential_6"
_____
Layer (type)                 Output Shape              Param #
=================================================================
dense_10 (Dense)             (None, 512)               401920
_____
dropout_5 (Dropout)          (None, 512)               0
_____
dense_11 (Dense)             (None, 10)                5130
=================================================================
Total params: 407,050
Trainable params: 407,050
Non-trainable params: 0
_____
```

还原的模型使用与原始模型相同的参数进行编译。尝试使用加载的模型运行评估和预测。

```
#评估还原的模型
loss, acc = new_model.evaluate(test_images,  test_labels, verbose=2)
print('Restored model, accuracy: {:5.2f}%'.format(100 * acc))
print(new_model.predict(test_images).shape)
32/32 - 0s - loss: 0.4630 - accuracy: 0.0890
Restored model, accuracy:  8.90%
(1000, 10)
```

还原后的模型分类准确率达到了 8.90%，同时由 save_model 保存的模型结果既保持了模型的图结构，又保存了模型的参数。可以使用 load_model 加载保存的 h5 模型文件，重新实例化模型，如果文件中存储了训练配置的话，该函数还会同时完成模型的编译。模型可以在训练期间和训练完成后进行保存。这意味着模型可以从任意中断中恢复，并避免在训练上耗费比较长的时间。

第 3 章

图像处理：增广与分类

数据集的处理涉及大规模的批量自动化处理。本章主要通过介绍图像数据集中的图像处理，让大家初步了解数据处理的基本方法，以期望大家对数据预处理可以提高深度学习训练水平能够有一个更直观的了解。本章主要介绍 TensorFlow 在图像处理中的应用，包括图像增广、图像分类、图像分割方面的实战。

3.1 数据增广之图像变换实战

本节将演示图像数据增广的一系列方法，通过对图像进行旋转、翻转、反色等操作得到一系列新的图像，扩充数据集多样性。本节将介绍 Keras 预处理层和 tf.image 两种方法来实现图像增广[9]。

3.1.1 背景原理

在实际工作中常常会遇到数据量太小，模型得不到充分训练的情况。在这种情况下可以使用数据增强对图像进行旋转、位移、反转、反色等操作，成倍数地增加数据集的数量。进行数据增广的理论前提是：一个欠训练的神经网络会将经过位移、旋转的图像认为是不同的图像，而一个经过训练具有不变性的神经网络可以稳健地对这些图像进行分类。数据增广在实际工作中可以有效扩充数据集数量，使模型得到充分的训练，是改善结果和避免过拟合的常用技术。

3.1.2 安装操作

本节所用代码文件结构如表 3-1 所示。

表 3-1 代码文件结构

文 件 名 称	实 现 功 能
data_augmentation.ipynb	包含所有图像增广操作

本次实战只有 data_augmentation.ipynb 这一个代码文件，里面包含所有的图像增广操作。安装步骤如下。

（1）进入 Anaconda Prompt 命令行，切换至本章根目录。
（2）在命令行中输入 jupyter notebook。
（3）在浏览器中打开 data_augmentation.ipynb。

(4) 分步运行代码，查看运行结果。

3.1.3 代码解析

1. 安装并导入 TensorFlow 和第三方库

```
import matplotlib.pyplot as plt
import numpy as np
import tensorflow as tf
import tensorflow_datasets as tfds
from tensorflow.keras import layers
from tensorflow.keras.datasets import mnist
```

2. 下载数据集

本节使用 tf_flowers 数据集，该数据集可以通过 tensorflow_datasets 下载获取，其中，前 80% 的数据用于训练，范围 80%～90% 内的数据用于验证，最后 10% 的数据用于测试。

```
(train_ds, val_ds, test_ds), metadata = tfds.load(
    'tf_flowers',
    split=['train[:80%]', 'train[80%:90%]', 'train[90%:]'],
    with_info=True,
    as_supervised=True,
)
```

从数据集中检索一幅图像，并用它来演示数据扩充。原始图像如图 3-1 所示。

```
get_label_name = metadata.features['label'].int2str
image, label = next(iter(train_ds))
_ = plt.imshow(image)
_ = plt.title(get_label_name(label))
```

图 3-1 原始图像

3. 使用 Keras 预处理层

调整大小和缩放。

可以使用预处理图层将图像调整为一致的形状,并重新调整像素值。图 3-2 是将原始图像重新调整为 180 像素×180 像素之后的结果。

```
IMG_SIZE = 180
resize_and_rescale = tf.keras.Sequential([
  layers.experimental.preprocessing.Resizing(IMG_SIZE, IMG_SIZE),
  layers.experimental.preprocessing.Rescaling(1./255)
])
```

放大单个图像,垂直或水平翻转图像。可以看到将这些图层应用于图像的结果。

```
result = resize_and_rescale(image)
_ = plt.imshow(result)
```

图 3-2　图像重新调整大小

如下面的代码和结果展示所示,可以验证像素在 0～1 中。

```
print("Min and max pixel values:", result.numpy().min(), result.numpy().max())
Min and max pixel values: 0.0 1.0
```

4. 数据增强

可以使用预处理层进行数据增强。创建一些预处理层,并将它们重复应用于同一图像。图 3-3 显示了 9 张图像,这 9 张图像是原始图像分别被随机旋转 9 次之后得到的图像。

```
data_augmentation = tf.keras.Sequential([
  layers.experimental.preprocessing.RandomFlip("horizontal_and_vertical"),
  layers.experimental.preprocessing.RandomRotation(0.2),
])
```

```
#将图像添加到批次
image = tf.expand_dims(image, 0)
    plt.figure(figsize=(10, 10))
for i in range(9):
  augmented_image = data_augmentation(image)
  ax = plt.subplot(3, 3, i + 1)
  plt.imshow(augmented_image[0])
  plt.axis("off")
```

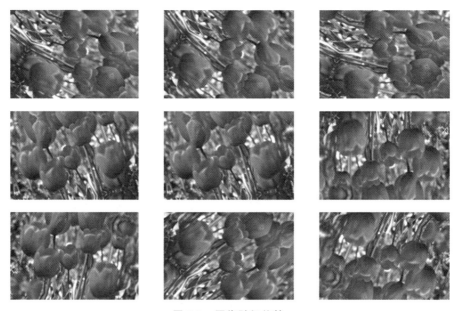

图 3-3 图像随机旋转

可以使用多种预处理层进行数据增强，包括 layer.RandomContrast、layers.RandomCrop、layers.RandomZoom 等。

5. 使用预处理层的两个选项

可以通过两种方式使用这些预处理层。

选项 1：使预处理层成为模型的一部分。

```
model = tf.keras.Sequential([
  resize_and_rescale,
  data_augmentation,
  layers.Conv2D(16, 3, padding='same', activation='relu'),
  layers.MaxPooling2D(),
  #Rest of your model
])
```

在这种情况下，有两点需要注意。
(1) 数据增强将与其余层同步在设备上运行，并受益于 GPU 加速。

（2）使用 model.save 导出模型时，预处理层将与模型的其余部分一起保存。如果以后部署此模型，它将自动标准化图像（根据图层的配置）。这样可以避免重新实现该逻辑服务器端的工作。

选项2：将预处理图层应用于数据集。

```
aug_ds = train_ds.map(
lambda x, y: (resize_and_rescale(x, training=True), y))
```

通过这种方法，可以使用 Dataset.map 创建一个生成一批增强图像的数据集。在此情况下：数据扩充将在 CPU 上异步发生，并且是非阻塞的。可以使用 Dataset.prefetch 将GPU 上的模型训练与数据预处理重叠。在这种情况下，当调用 model.save 时，预处理层将不会随模型一起导出。在保存模型或在服务器端重新实现它们之前，需要将它们附加到模型上。训练后可以在导出之前附加预处理层。可以在图像分类教程中找到第一个选项的示例。在这里演示第二个选项。

6．将预处理层应用于数据集

使用上面创建的预处理层配置训练数据库、验证数据库和测试数据集。对数据集进行数据增强以提高性能，同时使用并行读取的预取而非阻塞 I/O 的方式和缓冲来从磁盘产生批次，而不会阻塞 I/O。

```
batch_size = 32
AUTOTUNE = tf.data.experimental.AUTOTUNE
def prepare(ds, shuffle=False, augment=False):
  #调整大小并重新缩放所有数据集
  ds = ds.map(lambda x, y: (resize_and_rescale(x), y),
            num_parallel_calls=AUTOTUNE)
  if shuffle:
    ds = ds.shuffle(1000)
  #批处理所有数据集
  ds = ds.batch(batch_size)
  #仅在训练集上使用数据增强
  if augment:
    ds = ds.map(lambda x, y: (data_augmentation(x, training=True), y),
            num_parallel_calls=AUTOTUNE)
  #在所有数据集上使用缓冲预检
  return ds.prefetch(buffer_size=AUTOTUNE)
    train_ds = prepare(train_ds, shuffle=True, augment=True)
val_ds = prepare(val_ds)
test_ds = prepare(test_ds)
```

7．训练模型

为了完整起见，现在将使用这些数据集训练模型。该模型尚未针对准确性进行调整。下面的代码代表网络模型的神经网络层构成和训练过程中的准确率。

```python
model = tf.keras.Sequential([
  layers.Conv2D(16, 3, padding='same', activation='relu'),
  layers.MaxPooling2D(),
  layers.Conv2D(32, 3, padding='same', activation='relu'),
  layers.MaxPooling2D(),
  layers.Conv2D(64, 3, padding='same', activation='relu'),
  layers.MaxPooling2D(),
  layers.Flatten(),
  layers.Dense(128, activation='relu'),
  layers.Dense(num_classes)
])
model.compile(optimizer='adam',
              loss=tf.keras.losses.SparseCategoricalCrossentropy(from_logits=True),
              metrics=['accuracy'])
epochs=5
history = model.fit(
  train_ds,
  validation_data=val_ds,
  epochs=epochs
)
    loss, acc = model.evaluate(test_ds)
print("Accuracy", acc)

12/12 [==============================] - 1s 83ms/step - loss: 0.8226 - accuracy: 0.6567
Accuracy 0.6566757559776306
```

8. 自定义数据扩充

可以创建自定义数据扩展层。首先，创建一个 layers.Lambda 层，这是编写简洁代码的好方法。接下来，通过子类编写一个新层，从而提供更多控制权。根据一定的概率，两层都会随机反转图像中的颜色。图 3-4 显示了经过随机颜色反转得到的 9 张图片。

```python
def random_invert_img(x, p=0.5):
  if  tf.random.uniform([]) < p:
    x = (255-x)
  else:
    x
  return x
def random_invert(factor=0.5):
  return layers.Lambda(lambda x: random_invert_img(x, factor))
random_invert = random_invert()
plt.figure(figsize=(10, 10))
for i in range(9):
  augmented_image = random_invert(image)
  ax = plt.subplot(3, 3, i + 1)
```

```
plt.imshow(augmented_image[0].numpy().astype("uint8"))
plt.axis("off")
```

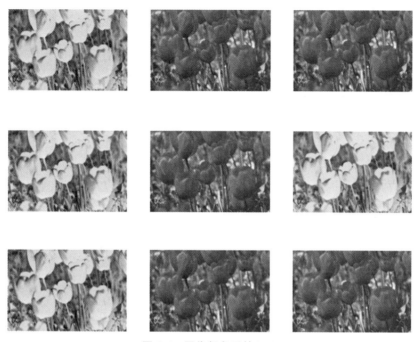

图 3-4　图像颜色反转（一）

接下来，通过子类实现自定义层。图 3-5 是一张经过两层处理后得到的图像。

```
class RandomInvert(layers.Layer):
  def __init__(self, factor=0.5, **kwargs):
    super().__init__(**kwargs)
    self.factor = factor
  def call(self, x):
    return random_invert_img(x)
_ = plt.imshow(RandomInvert()(image)[0])
```

图 3-5　图像颜色反转（二）

可以按照上文的选项 1 和选项 2 所述使用这两个层。

3.1.4 训练测试

1. 使用 tf.image

上文的 layers.preprocessing 实用程序很方便。为了更好地控制，可以使用 tf.data 和 tf.image 编写自己的数据增强管道或图层。

由于 flowers 数据集先前已配置了数据增强功能，因此将其重新导入以重新开始。

```
(train_ds, val_ds, test_ds), metadata = tfds.load(
    'tf_flowers',
    split=['train[:80%]', 'train[80%:90%]', 'train[90%:]'],
    with_info=True,
    as_supervised=True,
)
```

2. 检索要使用的图像

```
image, label = next(iter(train_ds))
_ = plt.imshow(image)
_ = plt.title(get_label_name(label))
```

使用以下功能来并排可视化和比较原始图像（见图 3-6）。

图 3-6　原始图像

```
def visualize(original, augmented):
fig = plt.figure()
plt.subplot(1,2,1)
plt.title('Original image')
plt.imshow(original)
plt.subplot(1,2,2)
plt.title('Augmented image')
plt.imshow(augmented)
```

3. 翻转图像

翻转图像指垂直或水平翻转图像(见图 3-7)。

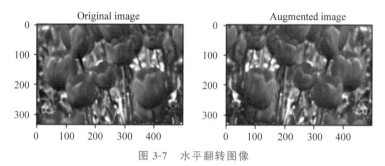

图 3-7　水平翻转图像

```
flipped = tf.image.flip_left_right(image)
visualize(image, flipped)
```

4. 灰度图像

产生如图 3-8 所示的灰度图像。

```
grayscaled = tf.image.rgb_to_grayscale(image)
visualize(image, tf.squeeze(grayscaled))
_ = plt.colorbar()
```

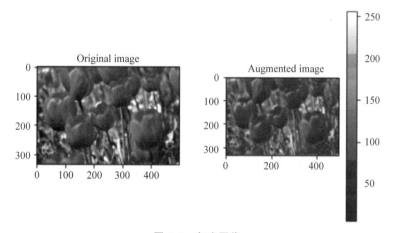

图 3-8　灰度图像

5. 饱和图像

通过提供饱和度因子使图像饱和(见图 3-9)。

```
saturated = tf.image.adjust_saturation(image, 3)
visualize(image, saturated)
```

图 3-9　图像饱和度变换

6. 更改图像亮度

通过提供亮度因子来更改图像的亮度（见图 3-10）。

```
bright = tf.image.adjust_brightness(image, 0.4)
visualize(image, bright)
```

图 3-10　图像亮度变换

7. 中心裁剪图像

从中心裁剪图像到所需的图像部分（见图 3-11）。

```
cropped = tf.image.central_crop(image, central_fraction=0.5)
visualize(image,cropped)
```

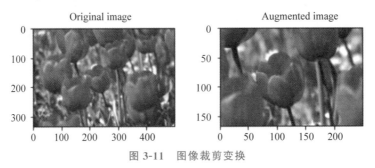

图 3-11　图像裁剪变换

8. 旋转图像

将图像旋转 90°（见图 3-12）。

```
rotated = tf.image.rot90(image)
visualize(image, rotated)
```

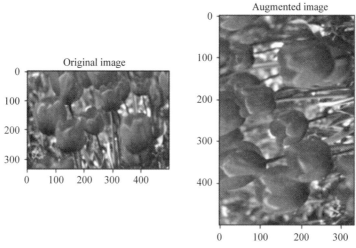

图 3-12　图像旋转 90°

9. 将增强应用于数据集

和之前一样，使用 Dataset.map 将数据扩充应用于数据集。

```
def resize_and_rescale(image, label):
    image = tf.cast(image, tf.float32)
    image = tf.image.resize(image, [IMG_SIZE, IMG_SIZE])
    image = (image / 255.0)
    return image, label
def augment(image,label):
    image, label = resize_and_rescale(image, label)
    #添加 6 像素的填充
    image = tf.image.resize_with_crop_or_pad(image, IMG_SIZE + 6, IMG_SIZE + 6)
    #随机裁剪回原始大小
    image = tf.image.random_crop(image, size=[IMG_SIZE, IMG_SIZE, 3])
    image = tf.image.random_brightness(image, max_delta=0.5)    #随机更改图像亮度
    image = tf.clip_by_value(image, 0, 1)
    return image, label
```

10. 配置数据集

```
train_ds = (
    train_ds
    .shuffle(1000)
    .map(augment, num_parallel_calls=AUTOTUNE)
    .batch(batch_size)
```

```
        .prefetch(AUTOTUNE)
)
val_ds = (
    val_ds
    .map(resize_and_rescale, num_parallel_calls=AUTOTUNE)
    .batch(batch_size)
    .prefetch(AUTOTUNE)
)
test_ds = (
    test_ds
    .map(resize_and_rescale, num_parallel_calls=AUTOTUNE)
    .batch(batch_size)
    .prefetch(AUTOTUNE)
)
```

这些数据集现在可以用于训练模型，如前所示。

3.2 在 CIFAR10 上应用 VGGNet 实现图像分类

本次实战将使用 VGGNet[10] 在 CIFAR10 数据集[11] 上进行图像分类。VGGNet 是一个深度神经网络，它探索了卷积神经网络的深度与其性能之间的关系。CIFAR10 是一个有 10 类图像的彩色图像分类数据集。此次实战将实现使用深度神经网络对多通道图像进行特征提取，实现图像分类。

3.2.1 背景原理

MNIST 是机器学习最常用的数据集之一，但由于手写数字图像非常简单，并且 MNIST 数据集只保存了图像灰度信息，并不适合输入设计为 RGB 通道的网络模型。本次实战将介绍另一个经典的图像分类数据集——CIFAR10。CIFAR10 数据集由加拿大 Canadian Institute For Advanced Research 发布，它包含了飞机、汽车、鸟、猫等共 10 类物体的彩色图像，每个种类收集了 6000 张 32 像素×32 像素大小的图像，共 6 万张。其中 5 万张作为训练数据集，1 万张作为测试数据集。

VGGNet 于 2014 年被提出，并于当年的 ILSVRC2014 比赛中取得了分类项目的第二名和定位项目的第一名。它通过构建深度神经网络，提升了网络的最终性能，使错误率大幅下降，同时拓展性与泛化性能也非常好。VGGNet 的网络结构包括 4 个阶段：输入、卷积层、全连接和输出层，如图 3-13 所示（详见彩插）。

其中图 3-13 的卷积层包括 5 个子模块，每个模块由 2 个或 3 个卷积层加上一个池化层组成。卷积层之后有 3 个全连接层，在最后接入一个 softmax 层进行分类，输出结果。

3.2.2 安装操作

本节所用代码文件结构如表 3-2 所示。

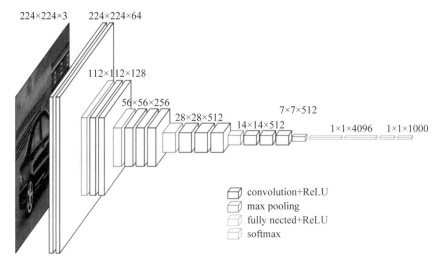

图 3-13　VGGNet 的网络结构

表 3-2　代码文件结构

文 件 名 称	实 现 功 能	文 件 名 称	实 现 功 能
main.py	数据预处理，训练、测试模型	network.py	定义 VGGNet 网络结构

本次实战的核心代码为 main.py。首先进入 Anaconda Prompt 命令行，在 main.py 所在目录下输入命令 python main.py，运行 main.py 文件即可得到测试结果。

在 TensorFlow 中，不需要手动下载、解析 CIFAR10 数据集，TensorFlow 会自动将数据集下载在"C:\Users\用户名\.keras\datasets"路径下，用户可以查看，也可以手动删除不需要的数据集缓存。

3.2.3　代码解析

VGGNet 原有的输入是 224 像素×224 像素的图像，而 CIFAR10 数据集中保存的图像像素是 32 像素×32 像素，因此需要修改网络结构以适应数据集输入。

在 network.py 文件中定义了 VGGNet 的网络模型，其中输入层的尺寸为 32 像素×32 像素×3，图像尺寸是 32 像素×32 像素，共有 RGB 3 个通道。卷积层的 5 个子模块基本结构代码如下所示。这是一个包含 2 个卷积层和一个池化层的基本模块，卷积层的激活函数为 ReLU，并使用 dropout 的方法来避免过拟合。

```
model.add(layers.Conv2D(64, (3, 3), padding='same',
input_shape=input_shape, kernel_regularizer=regularizers.l2(weight_decay)))
model.add(layers.Activation('relu'))
model.add(layers.BatchNormalization())
model.add(layers.Dropout(0.3))

model.add(layers.Conv2D(64, (3, 3),
padding='same',kernel_regularizer=regularizers.l2(weight_decay)))
model.add(layers.Activation('relu'))
```

```
model.add(layers.BatchNormalization())

model.add(layers.MaxPooling2D(pool_size=(2, 2)))
```

main.py 中实现了处理数据集以及训练的过程,核心代码如下所示,首先对数据集进行划分,之后将取值为 0~255 的像素值缩放至 0~1,并使用均值和方差进行归一化处理。

```
(x,y), (x_test, y_test) = datasets.cifar10.load_data()
x, x_test = normalize(x, x_test)
```

训练集的缓冲区大小为 50000,每一个批次(batch)的大小为 256。

```
train_loader = tf.data.Dataset.from_tensor_slices((x,y))
train_loader = train_loader.map(prepare_cifar).shuffle(50000).batch(256)
```

测试集的缓冲区大小为 10000,每一个批次(batch)的大小为 256。

```
test_loader = tf.data.Dataset.from_tensor_slices((x_test, y_test))
test_loader = test_loader.map(prepare_cifar).shuffle(10000).batch(256)
```

损失函数使用交叉熵,用准确率作为评判指标衡量模型性能并使用 Adam 优化器以 0.0001 的初始学习率进行优化。

3.2.4 训练测试

种类样片如图 3-14 所示。

图 3-14　种类样

运行 main.py 文件即可开始训练模型,在训练完 50 个 epoch 后,网络在测试集上的准确率达到了 75%,如图 3-15 所示。

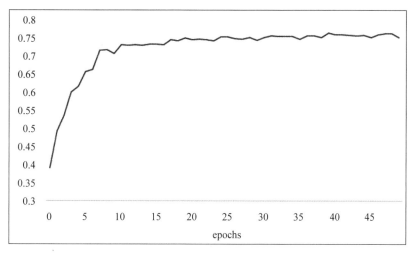

图 3-15　测试集准确率

3.3　图像识别之猫狗分类

本节主要介绍 TensorFlow 在图像识别方面的应用,可以对图像进行识别并分类。本次实战主要通过详细的代码实践完成神经网络层的讲解与编程,通过设定动物品种(猫、狗)并输入图像训练,然后运用卷积神经网络完成动物品种的识别。

3.3.1　背景原理

图像分类是深度学习中十分经典的一类任务,已经有很多深度学习的模型被提出并用于图像分类。实际上,如果不追求很高的分类准确度,只是使用神经网络来完成图像分类,那么一个很简单的神经网络就能够完成这项任务。

本次实战使用 Google 公司提供的猫狗数据集,搭建了一个很简单的只有 3 个卷积块的神经网络来完成这个简单的二分类问题。通过这个简单的例子,可以为之后更复杂的内容先打下基础。

3.3.2　安装操作

本节所用代码文件结构如表 3-3 所示。

表 3-3　代码文件结构

文 件 名 称	实 现 功 能
classification.ipynb	包含所有图像分类操作

本次实战只有 classification.ipynb 这一个代码文件,里面包含所有的图像分类操作。安装步骤如下。

(1) 进入 Anaconda Prompt 命令行,切换至本章根目录。
(2) 在命令行中输入 jupyter notebook。
(3) 在浏览器中打开 classification.ipynb。
(4) 分步运行代码,查看运行结果。

3.3.3 代码解析

1. 安装并导入 TensorFlow 和第三方库

```
import tensorflow as tf
from tensorflow.keras.models import Sequential
from tensorflow.keras.layers import Dense, Conv2D, Flatten, Dropout, MaxPooling2D
from tensorflow.keras.preprocessing.image import ImageDataGenerator
import os
import numpy as np
import matplotlib.pyplot as plt
```

2. 加载数据集

```
_URL = 'https://storage.googleapis.com/mledu-datasets/cats_and_dogs_filtered.zip'
path_to_zip = tf.keras.utils.get_file('cats_and_dogs.zip', origin=_URL, extract=True)
PATH = os.path.join(os.path.dirname(path_to_zip), 'cats_and_dogs_filtered')
```

在提取其内容之后,为训练集和验证集分配具有适当文件路径的变量。

```
train_dir = os.path.join(PATH, 'train')
validation_dir = os.path.join(PATH, 'validation')
train_cats_dir = os.path.join(train_dir, 'cats')
train_dogs_dir = os.path.join(train_dir, 'dogs')
validation_cats_dir = os.path.join(validation_dir, 'cats')
validation_dogs_dir = os.path.join(validation_dir, 'dogs')
```

3. 创建模型

该模型由 3 个卷积块组成,每个卷积块中都有一个最大池化层和一个二维卷积层。之后有一个全连接层,共有 512 个神经元,由 ReLU 激活函数激活。

```
model = Sequential([
    Conv2D(16, 3, padding='same', activation='relu', input_shape=(IMG_HEIGHT, IMG_WIDTH ,3)),
    MaxPooling2D(),
    Conv2D(32, 3, padding='same', activation='relu'),
    MaxPooling2D(),
    Conv2D(64, 3, padding='same', activation='relu'),
```

```
        MaxPooling2D(),
        Flatten(),
        Dense(512, activation='relu'),
        Dense(1)
])
```

4. 编译模型

选择 ADAM 优化器和二元交叉熵损失函数。

```
model.compile(optimizer='adam',
              loss=tf.keras.losses.BinaryCrossentropy(from_logits=True),
              metrics=['accuracy'])
```

3.3.4 训练测试

1. 训练模型

用类的方法训练网络。

```
history = model.fit_generator(
    train_data_gen,
    steps_per_epoch=total_train // batch_size,
    epochs=epochs,
    validation_data=val_data_gen,
    validation_steps=total_val // batch_size
)
```

2. 展示训练结果

```
acc = history.history['accuracy']
val_acc = history.history['val_accuracy']
loss=history.history['loss']
val_loss=history.history['val_loss']
epochs_range = range(epochs)
plt.figure(figsize=(8, 8))
plt.subplot(1, 2, 1)
plt.plot(epochs_range, acc, label='Training Accuracy')
plt.plot(epochs_range, val_acc, label='Validation Accuracy')
plt.legend(loc='lower right')
plt.title('Training and Validation Accuracy')
plt.subplot(1, 2, 2)
plt.plot(epochs_range, loss, label='Training Loss')
plt.plot(epochs_range, val_loss, label='Validation Loss')
plt.legend(loc='upper right')
plt.title('Training and Validation Loss')
plt.show()
```

结果图如图 3-16 所示。

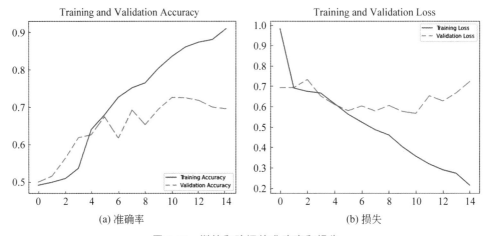

图 3-16　训练和验证的准确率和损失

从图 3-16 中可以看出，训练精度和验证精度相差很大，模型在验证集上的精度仅达到 70％左右。

第 4 章

图像增强识别实战

图像的修复与识别是计算机视觉中的重要任务之一。这些任务涉及了图像的特征初步提取与加工,并且对图像进行初步的修复和精化,这也是为更进一步的图像识别提供了基础。本章在第 3 章的基础上,进一步讨论了计算机视觉中最常见的几个任务:提高图像分辨率、识别图像色彩、使用注意力机制给图像取标题。

4.1 应用高阶神经网络提高图像分辨率

超分辨率技术(Super-Resolution,SR)是指从观测到的低分辨率图像重建出相应的高分辨率图像,较新的基于深度学习的 SR 方法,包括 SRCNN(超分辨率卷积网络)、VRSR(高分辨率网络)等。本次实战将介绍基于 TensorFlow 2.x 实现 EDSR、WDSR 和 SRGAN 这 3 种单幅图像超分辨率方法[12]。

4.1.1 背景原理

EDSR 采用增强型深度残差网络实现单幅图像超分辨率,是 NTIRE 2017 超分辨率挑战赛获得冠军的方案[13]。EDSR 最有意义的模型性能提升是去除掉了 SRResNet 多余的模块,从而可以扩大模型的尺寸来提升结果质量。

WDSR 是基于 CNN 的 SR 算法,其创新点在于扩大了 Block 中 ReLU 之前的特征图数量[14]。将大卷积核拆分成两个低级卷积核,节省参数,进一步在同参数的情况下提高 ReLU 前的特征图 channel 数。

SRGAN 将生成对抗网络(GAN)用于超分辨率问题,主要利用感知损失和对抗损失来提升恢复出的图像的真实感[15]。其中,感知损失是利用卷积神经网络提取出的特征,通过比较生成图像经过卷积神经网络后的特征和目标图像经过卷积神经网络后的特征的差别,使生成图像和目标图像在语义及风格上更相似。

4.1.2 安装操作

本节所用代码文件结构如表 4-1 所示。

表 4-1 代码文件结构

文 件 名 称	实 现 功 能	文 件 名 称	实 现 功 能
common.py	评估、归一化等方法	edsr.py	edsr 模型

续表

文 件 名 称	实 现 功 能	文 件 名 称	实 现 功 能
srgan.py	srgan 模型	wdsr.py	wdsr 模型
article.ipynb	实现原理	example-edsr.ipynb	edsr 模型训练和演示
example-srgan.ipynb	srgan 模型训练和演示	example-wdsr.ipynb	wdsr 模型训练和演示
train.py	定义训练过程	utils.py	加载和显示图像
data.py	准备数据集		

安装步骤如下。

（1）安装必要的 Python 包，目录可以在 environment.yml 中查看。

（2）下载预训练模型，放在文件根目录下面。

① Weight-EDSR-16-x4.tar.gz：EDSR x4。16 个残差块，64 个滤波器，参数达 1.52M。DIV2K 验证集上的 PSNR＝28.89dB（包含图像 801～900，6＋4 像素边界）。

② Weight-wdsr-b-32-x4.tar.gz：WDSR B x4。32 个残差块，32 个滤波器，参数达 0.62M。DIV2K 验证集上的 PSNR＝28.91dB（图像 801～900，包含 6＋4 像素边界）。

③ Weight-srgan.tar.gz：SRGAN。参数达 1.55M，使用 VGG54 content loss 进行训练。

（3）进入 Anaconda Prompt 命令行，切换至本章根目录，进入此次实战的目录。

（4）在命令行中输入 jupyter notebook。

（5）在浏览器中打开 article.ipynb、example-edsr.ipynb、example-srgan.ipynb、example-wdsr.ipynb 等几个文件。

（6）分步运行代码，查看运行结果。

4.1.3　代码解析

在实验中使用 DIV2K 数据集，其中有 1000 张高清图（2K 分辨率），这些图中 800 张用作训练，100 张用作验证，100 张作为测试。使用提供的 DIV2K 的 data loader 自动下载 DIV2K 图像到.div2k 目录，并将它们转换为不同的格式，以便更快地加载。训练数据集如下所示。

```
from data import DIV2K
train_loader = DIV2K(scale=4, downgrade='bicubic', subset='train')
train_ds = train_loader.dataset(batch_size=16, random_transform=True, repeat_count=None)
for lr, hr in train_ds:
    ...
```

因为 DIV2K 图像大小不完全相同，所以把 batch size 设置为 1。加载验证集如下所示。

```
from data import DIV2K
valid_loader = DIV2K(scale=4, downgrade='bicubic', subset='valid')
valid_ds = valid_loader.dataset(batch_size=1, random_transform=False,
```

```
                    repeat_count=1)
```

下一步,训练网络。

EDSR 的网络结构如下。

(1) 标准化输入。

(2) head:卷积×1。

(3) body:resblock(卷积×2+relu)×16。

(4) tail:卷积×($n+1$)。

(5) 标准化输出。

其上采样部分集成在 tail 的前 n 次卷积中,其中每次卷积将 channels 提升 4 倍,然后转化为放大 2 倍的 features(即将 4 个 channels 合并为一个长宽加倍的 channel),重复 $\log(n, 2)$ 次。在训练 EDSR 时,对 EDSR 模型进行 30 万步的训练,对 DIV2K 验证集的前 10 张图像每运行 1000 步就进行一次评估。并且仅当评估指标 PSNR 有所提高时,才保存一个 checkpoint。主要的训练代码如下所示。

```
from model.edsr import edsr
from train import EdsrTrainer
trainer = EdsrTrainer(model=edsr(scale=4, num_res_blocks=16),
                      checkpoint_dir=f'.ckpt/edsr-16-x4')
trainer.train(train_ds,valid_ds.take(10),steps=300000,evaluate_every=1000,
              save_best_only=True)
trainer.restore()
psnr = trainer.evaluate(valid_ds)
print(f'PSNR = {psnr.numpy():3f}')
trainer.model.save_weights('weights/edsr-16-x4/weights.h5')
```

WDSR_b 网络结构如下。

(1) 标准化输入。

(2) head:卷积×1。

(3) body:block(卷积+relu+卷积×2)×16。

(4) tail:卷积×1。

(5) skip:卷积×1。

(6) 标准化输出。

tail 部分仅有一次卷积,实际放大思路同 EDSR,不过是一次卷积得到足够的 channels (3 * n^2),一步重建为 3 通道的图像。WDSR 的训练代码如下所示。

```
from model.wdsr import wdsr_b
from train import WdsrTrainer
trainer = WdsrTrainer(model=wdsr_b(scale=4, num_res_blocks=32),
                      checkpoint_dir=f'.ckpt/wdsr-b-8-x4')
trainer.train(train_ds,valid_ds.take(10), steps=300000, evaluate_every=1000,
              save_best_only=True)
trainer.restore()
psnr = trainer.evaluate(valid_ds)
```

```
print(f'PSNR = {psnr.numpy():3f}')
trainer.model.save_weights('weights/wdsr-b-32-x4/weights.h5')
```

关于SRGAN的训练代码如下所示。

生成预训练模型代码。

```
from model.srgan import generator
from train import SrganGeneratorTrainer
SrganGeneratorTrainer(model=generator(),checkpoint_dir=f'.ckpt/pre_generator')
pre_trainer.train(train_ds,valid_ds.take(10),steps=1000000,evaluate_every=1000)
pre_trainer.model.save_weights('weights/srgan/pre_generator.h5')
```

生成对抗网络微调模型代码。

```
from model.srgan import generator, discriminator
from train import SrganTrainer
#创建一个generator 并用预训练的权重初始化该generator
gan_generator = generator()
gan_generator.load_weights('weights/srgan/pre_generator.h5')
#为GAN创建训练器
gan_trainer = SrganTrainer(generator=gan_generator, discriminator=discriminator())
#对GAN进行200000步的训练
gan_trainer.train(train_ds, steps=200000)
#保存generator和discriminator的权重
gan_trainer.generator.save_weights('weights/srgan/gan_generator.h5')
gan_trainer.discriminator.save_weights('weights/srgan/gan_discriminator.h5')
```

4.1.4　训练测试

EDSR：在example-edsr.ipynb中运行以下代码得到EDSR模型对图像进行超分辨率处理的结果。

```
from model import resolve_single
from model.edsr import edsr
from utils import load_image, plot_sample
model = edsr(scale=4, num_res_blocks=16)
model.load_weights('weights/edsr-16-x4/weights.h5')
lr = load_image('demo/0851x4-crop.png')
sr = resolve_single(model, lr)
plot_sample(lr, sr)
```

运行结果如图4-1所示。其中图4-1(a)是原图，图4-1(b)是超分辨率图像。

可以看到图4-1(b)的分辨率明显得到了提升，画面细节更加丰富，光晕效果更佳明显。

WDSR：在example-wdsr.ipynb中运行以下代码得到EDSR模型对图像进行超分辨率处理的结果。

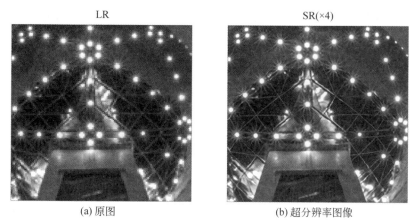

(a) 原图 (b) 超分辨率图像

图 4-1　EDSR 运行结果

```
from model.wdsr import wdsr_b
model = wdsr_b(scale=4, num_res_blocks=32)
model.load_weights('weights/wdsr-b-32-x4/weights.h5')
lr = load_image('demo/0829x4-crop.png')
sr = resolve_single(model, lr)
plot_sample(lr, sr)
```

运行结果如图 4-2 所示。其中,图 4-2(a)是原图,图 4-2(b)是超分辨率图像。

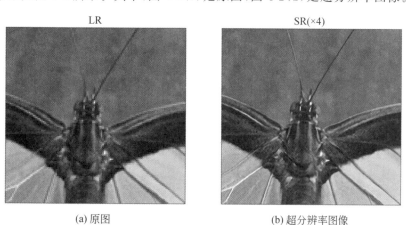

(a) 原图 (b) 超分辨率图像

图 4-2　WDSR 运行结果

图 4-2(b)中,图像更加清晰,昆虫的特征更加明显,纹理更清楚。

SRGAN:在 example-srgan.ipynb 中运行以下代码得到 EDSR 模型对图像进行超分辨率处理的结果。

```
from model.srgan import generator
model = generator()
model.load_weights('weights/srgan/gan_generator.h5')
lr = load_image('demo/0869x4-crop.png')
sr = resolve_single(model, lr)
plot_sample(lr, sr)
```

运行结果如图 4-3 所示。其中图 4-3(a)是原图,图 4-3(b)是超分辨率图像。

(a) 原图　　　　　　　　　　(b) 超分辨率图像

图 4-3　SRGAN 运行结果

图 4-3(b)中,画面更为清晰,细节更为丰富,与图 4-3(a)相比,猫的毛发更为清楚,根根分明,瞳孔边缘清楚。

4.2　长短期记忆机器人识别色彩

本次实战实现了一个简单的生成颜色的机器人,可以在给定颜色名称时可视化颜色。主要利用一个堆叠的 LSTM[16],它从颜色名称文本数据生成 RGB 颜色值,并将颜色显示出来。

4.2.1　背景原理

本次实战通过循环神经网络 RNN 实现预测生成颜色,而每个单元则采用长短期记忆的方式来具体实现。输入是一个元组(字符,序列长度)。其中,字符是一批一维张量表示的热编码颜色名称[批处理大小,时间步长,256],序列长度保持长度,每个字符序列(颜色名称)作为一个具有维度的张量[batch_size]。输出为:一种量纲张量[批量大小,标签尺寸],通过多层 RNN 传递字符,再应用 ReLU 达到最终隐藏状态。

为每个示例提取正确的输出(即隐藏状态)。每一批中的所有字符序列都被填到相同的固定长度,这样它们就可以很容易地通过上述 RNN 循环进行反馈。"序列长度"向量表示字符序列的真实长度,为每个序列获得由非填充字符生成的隐藏状态,从而得到最终的颜色预测结果。

4.2.2　安装操作

本节所用代码文件结构如表 4-2 所示。

表 4-2　代码文件结构

文 件 名 称	实 现 功 能	文 件 名 称	实 现 功 能
model.py	定义循环神经网络	main.py	实现训练和测试

安装步骤如下。

（1）进入 Anaconda Prompt 命令行，切换至本章根目录，进入此次实战的目录。
（2）在命令行中输入 python main.py。
（3）训练完成后输入表示颜色的单词，颜色机器人会生成 RGB 值以及图像。

4.2.3　代码解析

本次实战的核心代码为 main.py。

1. 导入必要的函数库

```
import  os, six, time
import  tensorflow as tf
import  numpy as np
from    tensorflow import keras
from    matplotlib import pyplot as plt
from    utils import load_dataset, parse
from    model import RNNColorbot
tf.random.set_seed(22)
np.random.seed(22)
os.environ['TF_CPP_MIN_LOG_LEVEL'] = '2'
assert tf.__version__.startswith('2.')
def test(model, eval_data):
```

2. 在验证数据集上计算平均损失率

```
avg_loss = keras.metrics.Mean()
for (labels, chars, sequence_length) in eval_data:
    predictions = model((chars, sequence_length), training=False)
    avg_loss.update_state(keras.losses.mean_squared_error(labels, predictions))
print("eval/loss: %.6f" % avg_loss.result().numpy())
def train_one_epoch(model, optimizer, train_data, log_interval, epoch):
```

3. 使用优化器在训练集上进行训练

```
for step, (labels, chars, sequence_length) in enumerate(train_data):
    with tf.GradientTape() as tape:
        predictions = model((chars, sequence_length), training=True)
        loss = keras.losses.mean_squared_error(labels, predictions)
        loss = tf.reduce_mean(loss)
    grads = tape.gradient(loss, model.trainable_variables)
    optimizer.apply_gradients(zip(grads, model.trainable_variables))
    if step % 100 == 0:
        print(epoch, step, 'loss:', float(loss))
SOURCE_TRAIN_URL = " https://raw.githubusercontent.com/random-forests/tensorflow-workshop/master/archive/extras/colorbot/data/train.csv"
```

```python
SOURCE_TEST_URL = "https://raw.githubusercontent.com/random-forests/
tensorflow-workshop/master/archive/extras/colorbot/data/test.csv"
def main():
    batchsz = 64
    rnn_cell_sizes = [256, 128]
    epochs = 40
    data_dir = os.path.join('.', "data")
    train_data = load_dataset(
        data_dir=data_dir, url=SOURCE_TRAIN_URL, batch_size=batchsz)
    eval_data = load_dataset(
        data_dir=data_dir, url=SOURCE_TEST_URL, batch_size=batchsz)
    model = RNNColorbot(
        rnn_cell_sizes=rnn_cell_sizes,
        label_dimension=3,
        keep_prob=0.5)
    optimizer = keras.optimizers.Adam(0.01)
    for epoch in range(epochs):
        start = time.time()
        train_one_epoch(model, optimizer, train_data, 50, epoch)
        end = time.time()
        #print("train/time for epoch #%d: %.2f" % (epoch, end - start))
        if epoch % 10 == 0:
            test(model, eval_data)
    print("Colorbot is ready to generate colors!")
    while True:
        try:
            color_name = six.moves.input("Give me a color name (or press enter to exit): ")
        except EOFError:
            return
        if not color_name:
            return
        _, chars, length = parse(color_name)
        (chars, length) = (tf.identity(chars), tf.identity(length))
        chars = tf.expand_dims(chars, 0)
        length = tf.expand_dims(length, 0)
        preds = tf.unstack(model((chars, length), training=False)[0])
        #预测值不可以为负值,但可以大于1

        clipped_preds = tuple(min(float(p), 1.0) for p in preds)
        rgb = tuple(int(p * 255) for p in clipped_preds)
        print("rgb:", rgb)
        data = [[clipped_preds]]
        plt.imshow(data)
        plt.title(color_name)
```

```
        plt.savefig(color_name+'.png')
if __name__ == "__main__":
    main()
```

4.2.4 训练测试

当训练完成时,每输入一个颜色,例如 blue(蓝色),就会得到相应颜色图像(见图 4-4)。

图 4-4　测试命令与效果

从图 4-4 中可以看到,通过训练使得颜色机器人能够通过给定的颜色生成对应的色彩图像,但在处理类似 dark 的单词时,颜色机器人看起来并不能正确生成对应的颜色。

4.3　使用注意力机制给图像取标题

给出一个类似图 4-5 所示的图像,目标是生成一个标题,例如"一个冲浪者骑在波浪上"。

图 4-5　给图像起标题

4.3.1　背景原理

要实现这一点,将使用基于注意力的模型,它能够在生成标题时看到模型关注图像的哪些部分(见图 4-6)。

本次实战是一个端到端的例子。运行时会下载 MS-COCO 数据集[17],使用 InceptionV3 预处理和缓存图像子集,训练编码器-解码器模型,并使用训练的模型在新图像上生成标题。在此示例中,将根据相对较少的数据训练模型——大约 2 万个图像的前 3 万个标题(因为数据集中每个图像都有多个标题)。

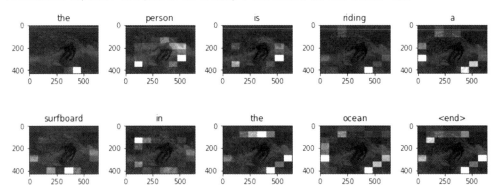

图 4-6 注意力机制

4.3.2 安装操作

本节所用代码文件结构如表 4-3 所示。

表 4-3 代码文件结构

文 件 名 称	实 现 功 能
image_captioning.ipynb	给图像做标题的核心文件

本次实战只有 image_captioning.ipynb 这一个代码文件，里面包含所有的图像增广操作。

安装步骤如下。

(1) 进入 Anaconda Prompt 命令行，切换至本章根目录。

(2) 在命令行中输入 jupyter notebook。

(3) 在浏览器中打开 image_captioning.ipynb。

(4) 分步运行代码，查看运行结果。

4.3.3 代码解析

1. 加载必要的函数库

```
import tensorflow as tf
import train_test_splitfrom sklearn.utils
import shuffleimport re
import numpy as np
import os
import time
import jsonfrom glob
import globfrom PIL
import Image
import pickle
```

2. 下载并准备 MS-COCO 数据集

将使用 MS-COCO 数据集训练模型。该数据集包含超过 8.2 万个图像,每个图像至少有 5 个不同的标题注释。下面的代码自动下载并提取数据集。

```
#下载标题注释
filesannotation_folder = '/annotations/'if not os.path.exists(os.path.abspath
('.') + annotation_folder): annotation_zip =tf.keras.utils.get_file('captions.
zip', cache_subdir=os.path.abspath('.'),
origin= 'http://images.cocodataset.org/annotations/annotations_trainval2014.
zip', extract = True)
annotation_file = os.path.dirname(annotation_zip) + '/annotations/captions_
train2014.json'  os.remove(annotation_zip)
#下载图像
filesimage_folder = '/train2014/'if not os.path.exists(os.path.abspath('.') +
image_folder):
image_zip = tf.keras.utils.get_file('train2014.zip', cache_subdir=os.path.
abspath('.'),
origin = 'http://images.cocodataset.org/zips/train2014.zip', extract = True)
PATH = os.path.dirname(image_zip) + image_folder  os.remove(image_zip)
else:   PATH = os.path.abspath('.') + image_folder
```

为了加快训练速度,将使用 3 万个标题及其对应图像的子集来训练模型。如果选择使用更多数据,生成标题的质量会得到提高。

```
#读取json文件
open(annotation_file, 'r') as f:
annotations = json.load(f)
#存储标题和图像名称
vectorsall_captions = []
all_img_name_vector = []
for annot in annotations['annotations']:
caption = '<start> ' + annot['caption'] + ' <end>'
image_id = annot['image_id']
full_coco_image_path = PATH + 'COCO_train2014_' + '%012d.jpg' % (image_id)   all
_img_name_vector.append(full_coco_image_path)   all_captions.append(caption)
#把标题和图像名称放在一起,并设定一个随机状态
train_captions, img_name_vector = shuffle(all_captions,  all_img_name_vector,
random_state=1)
#选择混合后的前30000个标题
setnum_examples = 30000
train_captions = train_captions[:num_examples]
img_name_vector = img_name_vector[:num_examples]
```

3. 使用 InceptionV3 对图像进行预处理

接下来,将使用 InceptionV3(在 Imagenet 上预先训练)对每个图像进行分类,将从最后

一个卷积层中提取特征。首先,通过以下方法将图像转换为InceptionV3[18]的预期格式:将图像大小调整为299像素×299像素(分别为图像的长度和宽度),使用PREPROCESS_INPUT方法对图像进行预处理,使其包含−1～1范围内的像素,这与训练InceptionV3所使用的图像格式相匹配。

```
def load_image(image_path):
    img = tf.io.read_file(image_path)
    img = tf.image.decode_jpeg(img, channels=3)
    img = tf.image.resize(img, (299, 299))
    img = tf.keras.applications.inception_v3.preprocess_input(img)
    return img, image_path
```

初始化InceptionV3并加载预先训练好的Imagenet权重。现在,将创建一个tf.keras模型,其中输出层是InceptionV3体系结构中的最后一个卷积层。该图层的输出形状为8×8×2048。

通过网络转换每个图像,并将生成的向量存储在字典中(image_name→Feature_Vector)。在所有图像都通过网络后,可以挑选字典并将其保存。

```
image_model = tf.keras.applications.InceptionV3(include_top=False, weights='imagenet')
new_input = image_model.input
hidden_layer = image_model.layers[-1].output
image_features_extract_model = tf.keras.Model(new_input, hidden_layer)
```

缓存从InceptionV3提取的特性,将使用InceptionV3对每个图像进行预处理,并将输出缓存到磁盘。在RAM中缓存输出会更快,但也会占用大量内存,每个图像需要8×8×2048个浮点数,如果机器内存只有4GB就会溢出。使用更复杂的缓存策略可以提高性能(例如,通过对图像进行分片以减少随机访问磁盘I/O),但这将需要更多代码。使用GPU运行缓存大约需要10min。

```
#获取预处理后的图像
encode_train = sorted(set(img_name_vector))
#随机更改batch_size的大小
configurationimage_dataset = tf.data.Dataset.from_tensor_slices(encode_train)
image_dataset = image_dataset.map(load_image, num_parallel_calls=tf.data.experimental.AUTOTUNE).batch(16)
for img, path in image_dataset:
    batch_features = image_features_extract_model(img)
    batch_features =tf.reshape(batch_features, (batch_features.shape[0], -1, batch_features.shape[3]))
    for bf, p in zip(batch_features, path):
        path_of_feature = p.numpy().decode("utf-8")
        np.save(path_of_feature, bf.numpy())
```

4. 预处理和标记化字幕

首先,将标记化字幕(例如,通过拆分空格)。这就提供了数据中所有唯一单词的词汇表

（例如"冲浪""足球"等）。接下来，将把词汇表大小限制为最大 5000 个单词（以节省内存），并将用标记 unk（未知）替换所有其他单词。然后创建词到索引和索引到词的映射。最后，将所有序列填充为与最长的序列长度相同。

```
#在数据集中查找最长的标题
def calc_max_length(tensor):
return max(len(t) for t in tensor)
#选择前 5000 个单词
from the vocabularytop_k = 5000
tokenizer =tf.keras.preprocessing.text.Tokenizer(num_words=top_k, oov_token=
"<unk>",filters= '!"#$%&()*+.,-/:;=?@[\]^_`{|}~ ')tokenizer.fit_on_texts
(train_captions)
train_seqs = tokenizer.texts_to_sequences(train_captions)
tokenizer.word_index['<pad>'] = 0tokenizer.index_word[0] = '<pad>'
#创建标签
vectorstrain_seqs = tokenizer.texts_to_sequences(train_captions)
#用向量的形式来表示最长标题
#如果没有提供最长标题的数值，那么 pad_sequences 函数会自动计算它的值
cap_vector = tf.keras.preprocessing.sequence.pad_sequences(train_seqs, padding
='post')
#计算用于最长标题的权重
weightsmax_length = calc_max_length(train_seqs)
```

5. 将数据拆分为训练集和测试集

```
#按照 80%和 20%的比例创建训练集和验证集
splitimg_name_train, img_name_val, cap_train, cap_val =train_test_split(img_
name_vector, cap_vector,  test_size=0.2, random_state=0)
len(img_name_train)
len(cap_train)
len(img_name_val)
len(cap_val)
(24000, 24000, 6000, 6000)
```

创建一个 tf.data 数据集用于训练，图像和字幕已经准备好了。接下来，创建一个 tf.data 数据集，用于训练模型。

```
#可根据需要随意更改以下参数
BATCH_SIZE = 64
BUFFER_SIZE = 1000
embedding_dim = 256
units = 512
vocab_size = top_k + 1
num_steps = len(img_name_train) // BATCH_SIZE
#从 InceptionV3 中提取的向量的形状为(64,2048)，用两个变量来表示该向量
shapefeatures_shape = 2048
```

```
attention_features_shape = 64
#加载 numpy 文件
def map_func(img_name, cap):
    img_tensor = np.load(img_name.decode('utf-8')+'.npy')
    return img_tensor, cap
dataset = tf.data.Dataset.from_tensor_slices((img_name_train, cap_train))
#使用 map 并行加载 numpy 文件
dataset = dataset.map(lambda item1, item2: tf.numpy_function(
        map_func, [item1, item2], [tf.float32, tf.int32]),
        num_parallel_calls=tf.data.experimental.AUTOTUNE)
#打乱顺序并进行批处理
dataset = dataset.shuffle(BUFFER_SIZE).batch(BATCH_SIZE)
dataset = dataset.prefetch(buffer_size=tf.data.experimental.AUTOTUNE)
```

模型：在本例中，从 InceptionV3 的较低卷积层提取特征，提供了一个形状为(8,8, 2048)的向量。随后把它压成(64,2048)的形状。此向量随后通过 CNN 编码器(由单个全连接层组成)。RNN(这里是 GRU)关注图像以预测下一个单词，具体的注意力引导机制详见下面的代码实现过程。

```
class BahdanauAttention(tf.keras.Model):
    def __init__(self, units):
        super(BahdanauAttention, self).__init__()
        self.W1 = tf.keras.layers.Dense(units)
        self.W2 = tf.keras.layers.Dense(units)
        self.V = tf.keras.layers.Dense(1)
    def call(self, features, hidden):
        #features(CNN_encoder output) shape == (batch_size, 64, embedding_dim)
        #hidden shape == (batch_size, hidden_size)
        #hidden_with_time_axis shape == (batch_size, 1, hidden_size)
        hidden_with_time_axis = tf.expand_dims(hidden, 1)
        #score shape == (batch_size, 64, hidden_size)
        score = tf.nn.tanh(self.W1(features) + self.W2(hidden_with_time_axis))
        #attention_weights shape == (batch_size, 64, 1)
        #you get 1 at the last axis because you are applying score to self.V
        attention_weights = tf.nn.softmax(self.V(score), axis=1)
        #context_vector shape after sum == (batch_size, hidden_size)
        context_vector = attention_weights * features
        context_vector = tf.reduce_sum(context_vector, axis=1)
        return context_vector, attention_weights
class CNN_Encoder(tf.keras.Model):
    #获取特征并用 pickle 将其序列化
    #这个编码器通过一个全连接层传递这些特性
    def __init__(self, embedding_dim):
        super(CNN_Encoder, self).__init__()
        #shape after fc == (batch_size, 64, embedding_dim)
```

```python
        self.fc = tf.keras.layers.Dense(embedding_dim)
    def call(self, x):
        x = self.fc(x)
        x = tf.nn.relu(x)
        return x
class RNN_Decoder(tf.keras.Model):
    def __init__(self, embedding_dim, units, vocab_size):
        super(RNN_Decoder, self).__init__()    self.units = units
    self.embedding = tf.keras.layers.Embedding(vocab_size, embedding_dim)
    self.gru =tf.keras.layers.GRU(self.units, return_sequences=True, return_state
=True, recurrent_initializer='glorot_uniform')
    self.fc1 = tf.keras.layers.Dense(self.units)
    self.fc2 = tf.keras.layers.Dense(vocab_size)
    self.attention = BahdanauAttention(self.units)
    def call(self, x, features, hidden):
#将注意力定义为一个单独的模型
    context_vector, attention_weights = self.attention(features, hidden)
#x shape after passing through
    embedding == (batch_size, 1, embedding_dim)
    x = self.embedding(x)
#x shape after concatenation == (batch_size, 1, embedding_dim + hidden_size)
    x = tf.concat([tf.expand_dims(context_vector, 1), x], axis=-1)
#将连接的向量传递给 GRU
    output, state = self.gru(x)
#shape == (batch_size, max_length, hidden_size)
    x = self.fc1(output)
#x shape == (batch_size * max_length, hidden_size)
    x = tf.reshape(x, (-1, x.shape[2]))
#output shape == (batch_size * max_length, vocab)
    x = self.fc2(x)
    return x, state, attention_weights
    def reset_state(self, batch_size):
    return tf.zeros((batch_size, self.units))
encoder = CNN_Encoder(embedding_dim)
decoder = RNN_Decoder(embedding_dim, units, vocab_size)
optimizer = tf.keras.optimizers.Adam()
loss_object = tf.keras.losses.SparseCategoricalCrossentropy(from_logits=True,
reduction='none')
def loss_function(real, pred):
mask = tf.math.logical_not(tf.math.equal(real, 0))
loss_ = loss_object(real, pred)
mask = tf.cast(mask, dtype=loss_.dtype)
loss_ *= mask    return tf.reduce_mean(loss_)
```

6. 保存训练好的模型

```
checkpoint_path = "./checkpoints/train" ckpt = tf.train.Checkpoint(encoder = encoder, decoder=decoder, optimizer = optimizer)
ckpt_manager =tf.train.CheckpointManager(ckpt, checkpoint_path, max_to_keep=5)
start_epoch = 0if ckpt_manager.latest_checkpoint:
start_epoch = int(ckpt_manager.latest_checkpoint.split('-')[-1])
#保存最新检查点的路径于 checkpoint_path
ckpt.restore(ckpt_manager.latest_checkpoint)
```

4.3.4 训练测试

提取存储在各自.npy 文件中的特征，然后通过解码器传递这些特征。
（1）将编码器输出、隐藏状态（初始化为0）和解码器输入（即开始令牌）传递给解码器。
（2）解码器返回预测和解码器隐藏状态。
（3）将解码器隐藏状态传递回模型，并使用预测值来计算损失。
（4）使用注意力机制决定解码器的下一个输入。
（5）计算梯度，并将其应用于优化器和反向传播。

```
#将其添加到单独的单元格中,因为如果多次运行训练单元格,则将重置 loss_plot 数组
loss_plot = []
@tf.function
def train_step(img_tensor, target):
  loss = 0
  #初始化每个批处理的隐藏状态
  hidden = decoder.reset_state(batch_size=target.shape[0])
  dec_input = tf.expand_dims([tokenizer.word_index['<start>']] * target.shape[0], 1)
  with tf.GradientTape() as tape:
      features = encoder(img_tensor)
      for i in range(1, target.shape[1]):
          #通过解码器传递
          predictions, hidden, _ = decoder(dec_input, features, hidden)
          loss += loss_function(target[:, i], predictions)

          dec_input = tf.expand_dims(target[:, i], 1)
  total_loss = (loss / int(target.shape[1]))
  trainable_variables = encoder.trainable_variables + decoder.trainable_variables
  gradients = tape.gradient(loss, trainable_variables)
  optimizer.apply_gradients(zip(gradients, trainable_variables))
  return loss, total_loss
EPOCHS = 20for epoch in range(start_epoch, EPOCHS):
start = time.time()
total_loss = 0
```

```
for (batch, (img_tensor, target)) in enumerate(dataset):
  batch_loss, t_loss = train_step(img_tensor, target)
  total_loss += t_loss
  if batch % 100 == 0:
    print ('Epoch {} Batch {} Loss {:.4f}'.format(epoch + 1, batch, batch_loss.numpy()
  / int(target.shape[1])))
  #存储并绘制损失函数随迭代次数变化图像
  loss_plot.append(total_loss / num_steps)
  if epoch % 5 == 0:
    ckpt_manager.save()
  print ('Epoch {} Loss {:.6f}'.format(epoch +1, total_loss/num_steps))
  print ('Time taken for 1 epoch {} sec\n'.format(time.time() - start))
  plt.plot(loss_plot)
  plt.xlabel('Epochs')
  plt.ylabel('Loss')
  plt.title('Loss Plot')
  plt.show()
```

损失随着迭代增加而改变,如图 4-7 所示。

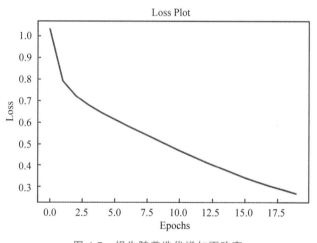

图 4-7　损失随着迭代增加而改变

EVALUE 函数类似于训练循环,不同之处在于这里不使用注意力机制。解码器在每个时间步长的输入是其先前的预测以及隐藏状态和编码器输出。当模型预测结束令牌时,停止预测,并存储每个时间步的注意力权重。

```
def evaluate(image):
  attention_plot = np.zeros((max_length, attention_features_shape))
  hidden = decoder.reset_state(batch_size=1)
  temp_input = tf.expand_dims(load_image(image)[0], 0)
  img_tensor_val = image_features_extract_model(temp_input)
  img_tensor_val = tf.reshape(img_tensor_val, (img_tensor_val.shape[0], -1,
img_tensor_val.shape[3]))
```

```
        features = encoder(img_tensor_val)
        dec_input = tf.expand_dims([tokenizer.word_index['<start>']], 0)
        result = []
        for i in range(max_length):
             predictions, hidden, attention_weights = decoder(dec_input, features, hidden)
            attention_plot[i] = tf.reshape(attention_weights, (-1, )).numpy()
            predicted_id = tf.random.categorical(predictions, 1)[0][0].numpy()
            result.append(tokenizer.index_word[predicted_id])
            if tokenizer.index_word[predicted_id] == '<end>':
                return result, attention_plot
            dec_input = tf.expand_dims([predicted_id], 0)
        attention_plot = attention_plot[:len(result), :]
        return result, attention_plot
def plot_attention(image, result, attention_plot):
    temp_image = np.array(Image.open(image))
    fig = plt.figure(figsize=(10, 10))
    len_result = len(result)
    for l in range(len_result):
        temp_att = np.resize(attention_plot[l], (8, 8))
        ax = fig.add_subplot(len_result//2, len_result//2, l+1)
        ax.set_title(result[l])
        img = ax.imshow(temp_image)
        ax.imshow(temp_att, cmap='gray', alpha=0.6, extent=img.get_extent())
    plt.tight_layout()
    plt.show()
```

运行下面这段代码，从数据集中随机选择一张图像得到图像的预测描述，并进行可视化输出。

```
rid = np.random.randint(0, len(img_name_val))
image = img_name_val[rid]
real_caption = ' '.join([tokenizer.index_word[i]
for i in cap_val[rid] if i not in [0]])
result, attention_plot = evaluate(image)
print ('Real Caption:', real_caption)
print ('Prediction Caption:', ' '.join(result))
plot_attention(image, result, attention_plot)
Real Caption: <start> a red double decker bus parked on a street in a european village <end>
Prediction Caption: the double deckered bus <unk> pulling trees <end>
```

真实描述：一辆红色双层巴士停在一个欧洲村庄的街道上。

预测描述：双层巴士拉着树。

图 4-8 说明了感知位置不同得到的预测文本结果也是不一样的，通过综合各个感知位置可以得到整张图像的标题。

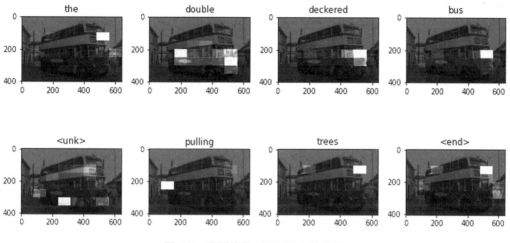

图 4-8 预测结果:双层巴士拉着树

第 5 章

两阶段目标检测实战

在 CNN 还未提出之前,针对目标检测这个问题,一般使用"设计受用特征+分类器"这个思路。当卷积神经网络一系列的模型被提出后,如 ResNet、GoogleNet、AlexNet 等,图像中的特征能够很好地提取出来。因此,目标检测的检测精确度得到极大的提高[19]。到后来推出了一系列的目标检测模型,如 RCNN[20]、Faster RCNN[21]、RPN[22]、MTCNN[23]等,进一步提升了检测精确度。本章主要介绍如何在 TensorFlow2 环境上实现基于 RPN、MTCNN 的两阶段目标检测。

5.1 基于 RPN 实现目标检测

本项目实现了 RPN(Region Proposal Network)的目标检测方法,RPN 区域生成网络作为双阶段目标检测中候选框的提取网络,可以加速检测过程。例如实际应用中,可以将 RPN 网络和 Fast RCNN(与 Faster RCNN 不一样,两种网络)网络结合到一起,将 RPN 获取到的输出框直接连到 ROI 池化层,实现一个 CNN 网络实现端到端目标检测的框架。

本节主要通过 Unity3D 产生虚拟数据,然后用 RPN 进行训练,再在真实数据上进行微调,实现行人检测实战,并利用 RPN 提高检测鲁棒性。

5.1.1 背景原理

RPN 候选框提取网络的输入输出如下。

输入:feature map、物体标签,即训练集中所有物体的类别与边框位置。

输出:Proposal、分类 Loss、回归 Loss。其中,Proposal 作为生成的区域,供后续模块分类与回归。两部分损失用作优化网络。

RPN 主要包含 3 部分:首先,生成 anchor boxes;其次,判断每个 anchor box 为前景(包含物体)或者后景(背景);最后,二分类边界框回归对 anchor box 进行微调,使得 positive anchor 和真实框(Ground Truth Box)更加接近。

在实现中,默认在每一个点上抽取 9 种 anchor,具体 Scale 为(8,1,32),Ratio 为(0.5,1,2),将这 9 种 anchor 的大小反算到原图上,即得到不同的原始 Proposal。然后通过分类网络与回归网络得到每一个 anchor 的前景背景概率和偏移量,回归网络将预测偏移量作用到 anchor 上使得 anchor 更接近于真实物体的真实坐标。参数选择中,在 feature map 上用 3×3 的卷积进行更深的特征提取。

在分类网络分支中,首先使用 1×1 卷积输出 18×37×50 的特征,由于每个点默认有 9 个 anchor,并且对每个 anchor 值预测其属于前景还是背景,因此通道数为 18。随后利用函

数将特征映射到 $2\times333\times75$，这样第一维仅仅是一个 anchor 的前景背景得分，并送到 Softmax 函数中进行概率计算，得到的特征再变换到 $18\times37\times50$ 的维度，最终输出的是每个 anchor 属于前景与背景的概率。

在回归分支中，利用 1×1 卷积输出 $36\times37\times50$ 的特征，第一位的 36 包含 9 个 anchor 的预测，每一个 anchor 有 4 个数据，分别代表了每一个 anchor 的中心点横、纵坐标及宽、高 4 个量相对于真值的偏移量。RPN 网络结构如图 5-1 所示。

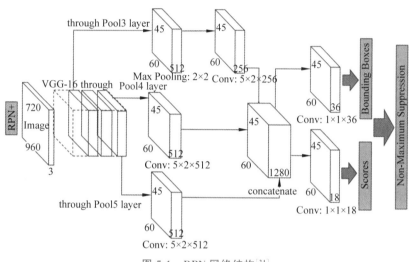

图 5-1　RPN 网络结构[24]

5.1.2　安装操作

本节所用代码文件结构如表 5-1 所示。

表 5-1　代码文件结构

文 件 名 称	实 现 功 能	文 件 名 称	实 现 功 能
kmeans.py	采用 k 均值算法生成 anchor	rpn.py	实现 RPN 模型网络结构
test.py	测试 RPN 模型	train.py	训练 RPN 模型
synthetic_dataset	合成行人数据集	utils.py	实现检测行人的工具函数，包括对 IOU 的计算等

（1）为了训练 RPN，需要使用合成数据集，它包含一组 8239 幅图像，其中只有一个类别（行人）。

（2）在训练过程中，如果候选边界框与真实框的交集和重叠度超过 50%，则将其视为正值；如果边界框与真实框的交集和重叠度小于 10%，则将被视为负值。运行一个 demo：python demo.py。

实验中采用 k 均值算法生成 9 个 anchor。可以运行 python kmeans.py 来查看生成过程。

（3）对网络进行训练，命令为 python train.py。

模型将在每个 epoch 中自动保存权重文件"./RPN.h5"。在神经网络训练期间，可以在

Tensorboard 中访问"http://localhost:6006/"跟踪损失函数曲线。

```
=> epoch 1   step 1   total_loss: 0.402951   score_loss: 0.346327   boxes_loss: 0.056625
=> epoch 1   step 2   total_loss: 0.399650   score_loss: 0.344363   boxes_loss: 0.055287
...
=> epoch 10   step 4000   total_loss: 0.001989   score_loss: 0.000015   boxes_loss: 0.001973
```

（4）用 200 张图像对模型进行测试，命令为 python test.py，预测结果存储在"./prediction"中。

5.1.3 代码解析

在 kmeans.py 中，使用 k 均值聚类方法求 9 个 anchor。通常，bounding box 由左上角顶点和右下角顶点表示。在对 box 做聚类时，只需要 box 的宽和高作为特征，并且由于数据集中图像的大小可能不同，还需要先使用图像的宽和高对 box 的宽和高做归一化。但如果直接使用标准 k 均值聚类中的欧几里得距离作为度量，则会有个问题，就是在聚类结果中，大 box 簇会比小 box 簇产生更大的误差（squared error）。由于只关心 anchor 与 box 的 IOU，不关心 box 的大小，因此，使用 IOU 作为度量更加合适。当 box 与 anchor 完全重叠，即 IOU=1 时，它们之间的距离为 0。因此，首先计算 box 与 anchor 之间的 IOU。

```
def iou(box, clusters):
    x = np.minimum(clusters[:, 0], box[0])
    y = np.minimum(clusters[:, 1], box[1])
    intersection = x * y
    box_area = box[0] * box[1]
    cluster_area = clusters[:, 0] * clusters[:, 1]
    iou_ = intersection / (box_area + cluster_area - intersection)
    return iou_
```

接着，对 box 进行 k 均值聚类，步骤如下。
（1）随机选取 k 个 box 作为初始 anchor。
（2）再使用 IOU 度量，将每个 box 分配给与其距离最近的 anchor。
（3）计算每个簇中所有 box 宽和高的均值，更新 anchor。
重复（2）、（3）步，直到 anchor 不再变化，或者达到了最大迭代次数。

```
def kmeans(boxes, k, dist=np.median, seed=1):
    rows = boxes.shape[0]
    distances = np.empty((rows, k))   ##N 行 x N 列
    last_clusters = np.zeros((rows,))
    np.random.seed(seed)
    #随机选取 k 个 box 作为初始 anchor
    clusters = boxes[np.random.choice(rows, k, replace=False)]
    while True:
```

```
        for icluster in range(k):
            distances[:,icluster] = 1 - iou(clusters[icluster], boxes)
        nearest_clusters = np.argmin(distances, axis=1)
        if (last_clusters == nearest_clusters).all():
            break
        for cluster in range(k):
            clusters[cluster] = dist(boxes[nearest_clusters == cluster], axis=0)
        last_clusters = nearest_clusters
    return clusters
```

在训练 train.py 中，encode_label() 完成从 groudtruth 的 box 坐标（x1，y1，x2，y2）到 regression deltas（dx，dy，dw，dh）的转换。

```
def encode_label(gt_boxes):
    target_scores = np.zeros(shape=[45, 60, 9, 2])     #0: 背景。1: 前景,
    target_bboxes = np.zeros(shape=[45, 60, 9, 4])     #t_x, t_y, t_w, t_h
    target_masks  = np.zeros(shape=[45, 60, 9])        #负样本: -1。正样本: 1
    for i in range(45):                                 #y: 高度
        for j in range(60):                             #x: 宽度
            for k in range(9):
                center_x = j * grid_width + grid_width * 0.5
                center_y = i * grid_height + grid_height * 0.5
                xmin = center_x - wandhG[k][0] * 0.5
                ymin = center_y - wandhG[k][1] * 0.5
                xmax = center_x + wandhG[k][0] * 0.5
                ymax = center_y + wandhG[k][1] * 0.5
                #忽略跨边界 anchor
                if (xmin > -5) & (ymin > -5) & (xmax < (image_width+5)) & (ymax < (image_height+5)):
                    anchor_boxes = np.array([xmin, ymin, xmax, ymax])
                    anchor_boxes = np.expand_dims(anchor_boxes, axis=0)
                    #计算此 anchor 和图像中所有的真实框之间的 IOU
                    ious = compute_iou(anchor_boxes, gt_boxes)
                    positive_masks = ious >= pos_thresh
                    negative_masks = ious <= neg_thresh
                    if np.any(positive_masks):
                        target_scores[i, j, k, 1] = 1.
                        target_masks[i, j, k] = 1       #分类为正样本
                        #找出匹配这个 anchor 的真实框
                        max_iou_idx = np.argmax(ious)
                        selected_gt_boxes = gt_boxes[max_iou_idx]
                        target_bboxes[i, j, k] = compute_regression(selected_gt_boxes, anchor_boxes[0])
                    if np.all(negative_masks):
                        target_scores[i, j, k, 0] = 1.
                        target_masks[i, j, k] = -1      #分类为负样本
```

```
return target_scores, target_bboxes, target_masks
```

计算损失函数：RPN 训练时要把 RPN classification 和 RPN bounding box regression 的 loss 加到一起来实现联合训练。loss 计算如下。

```
def compute_loss(target_scores, target_bboxes, target_masks, pred_scores, pred_
bboxes):
    score_loss=tf.nn.softmax_cross_entropy_with_logits(labels=target_scores,
 logits=pred_scores)
    foreground_background_mask = (np.abs(target_masks) == 1).astype(np.int)
    score_loss = tf.reduce_sum(score_loss * foreground_background_mask, axis=
[1,2,3]) / np.sum(foreground_background_mask)
    score_loss = tf.reduce_mean(score_loss)
    boxes_loss = tf.abs(target_bboxes - pred_bboxes)
    boxes_loss = 0.5 * tf.pow(boxes_loss, 2) * tf.cast(boxes_loss<1, tf.float32)
 + (boxes_loss - 0.5) * tf.cast(boxes_loss >=1, tf.float32)
    boxes_loss = tf.reduce_sum(boxes_loss, axis=-1)
    foreground_mask = (target_masks > 0).astype(np.float32)
    boxes_loss = tf.reduce_sum(boxes_loss * foreground_mask, axis=[1,2,3]) /
 np.sum(foreground_mask)
    boxes_loss = tf.reduce_mean(boxes_loss)
    return score_loss, boxes_loss
```

以下为核心代码 test.py，在测试时，首先对输出进行编码。

```
for i in range(45):
    for j in range(60):
        for k in range(9):
            center_x = j * grid_width + grid_width * 0.5
            center_y = i * grid_height + grid_height * 0.5
            xmin = center_x - wandhG[k][0] * 0.5
            ymin = center_y - wandhG[k][1] * 0.5
            xmax = center_x + wandhG[k][0] * 0.5
            ymax = center_y + wandhG[k][1] * 0.5
            #忽略跨边界的 anchor
            if (xmin > -5) & (ymin > -5) & (xmax < (image_width+5)) & (ymax <
 (image_height+5)):
                anchor_boxes = np.array([xmin, ymin, xmax, ymax])
                anchor_boxes = np.expand_dims(anchor_boxes, axis=0)
                #计算此 anchor 和图像中所有的真实框之间的 IOU
                ious = compute_iou(anchor_boxes, gt_boxes)
                positive_masks = ious > pos_thresh
                negative_masks = ious < neg_thresh
                if np.any(positive_masks):
                    plot_boxes_on_image(encoded_image, anchor_boxes, thickness=1)
                    print("=> Encoding positive sample: %d, %d, %d" %(i, j, k))
```

```python
                cv2.circle(encoded_image,center=(int(0.5*(xmin+xmax)),
                    int(0.5*(ymin+ymax))),radius=1,color=[255,0,0],thickness=4)
                target_scores[i, j, k, 1] = 1.
                target_masks[i, j, k] = 1         #设置标签为负样本
                #找出匹配这个 anchor 的真实框
                max_iou_idx = np.argmax(ious)
                selected_gt_boxes = gt_boxes[max_iou_idx]
                target_bboxes[i, j, k] = compute_regression(selected_gt_boxes,
            anchor_boxes[0])
            if np.all(negative_masks):
                target_scores[i, j, k, 0] = 1.
                target_masks[i, j, k] = -1        #设置标签为负样本
                cv2.circle(encoded_image,center=(int(0.5*(xmin+xmax)),
                    int(0.5*(ymin+ymax))),radius=1,color=[0,0,0],thickness=4)
    Image.fromarray(encoded_image).show()
```

再对输出进行解码。

```python
decode_image = np.copy(raw_image)
pred_boxes = []
pred_score = []
for i in range(45):
    for j in range(60):
        for k in range(9):
            #预测框坐标
            center_x = j * grid_width + 0.5 * grid_width
            center_y = i * grid_height + 0.5 * grid_height
            anchor_xmin = center_x - 0.5 * wandhG[k, 0]
            anchor_ymin = center_y - 0.5 * wandhG[k, 1]
            xmin = target_bboxes[i, j, k, 0] * wandhG[k, 0] + anchor_xmin
            ymin = target_bboxes[i, j, k, 1] * wandhG[k, 1] + anchor_ymin
            xmax = tf.exp(target_bboxes[i, j, k, 2]) * wandhG[k, 0] + xmin
            ymax = tf.exp(target_bboxes[i, j, k, 3]) * wandhG[k, 1] + ymin
            if target_scores[i, j, k, 1] > 0:      #正样本
                print("=> Decoding positive sample: %d, %d, %d" %(i, j, k))
                cv2.circle(decode_image,center=(int(0.5*(xmin+xmax)),int(0.5
*(ymin+ymax))),
                            radius=1, color=[255,0,0], thickness=4)
                pred_boxes.append(np.array([xmin, ymin, xmax, ymax]))
                pred_score.append(target_scores[i, j, k, 1])
pred_boxes = np.array(pred_boxes)
plot_boxes_on_image(decode_image, pred_boxes, color=[0, 255, 0])
Image.fromarray(np.uint8(decode_image)).show()
```

进行快速编码输出。

```python
faster_decode_image = np.copy(raw_image)
```

```
pred_bboxes = np.expand_dims(target_bboxes, 0).astype(np.float32)
pred_scores = np.expand_dims(target_scores, 0).astype(np.float32)
pred_scores, pred_bboxes = decode_output(pred_bboxes, pred_scores)
plot_boxes_on_image(faster_decode_image, pred_bboxes, color=[255, 0, 0]) Image.
fromarray(np.uint8(faster_decode_image)).show()
```

5.1.4 训练测试

从图 5-2(a)可以看到，RPN 准确地将识别出的人用方框框了出来。从图 5-2(b)可以看到，在目标的中心位置用了一个点来锚定物体的位置。

(a) 用方框框出识别出的人　　　　　　(b) 用一个点来锚定物体的位置

图 5-2　运行命令与效果展示

5.2　应用 MTCNN 实现人脸目标检测

本次实战将使用多任务卷积神经网络（Multi-Task Convolutional Neural Network，MTCNN）专门针对人脸进行目标检测，并且同时对人面部的 5 个标志点位（左眼、右眼、鼻尖、左右嘴角）进行目标检测。将人脸检测与人脸关键点对齐相结合是 MTCNN 的一大创新点，提高了人脸检测的性能。但同时 MTCNN 也存在着训练需要大量数据、训练时间长以及收敛速度慢等问题。

5.2.1　背景原理

MTCNN 是一个实现人脸识别和面部关键点检测的神经网络，它包括 P-Net、R-Net 和 O-Net 3 层网络结构[25]。

P-Net、R-Net 和 O-Net 是 3 个级联的网络，MTCNN 使用这 3 个网络进行逐层精化，一步步得到更好的预测结果。

P-Net（Proposal Network，提案网络）实现了人脸的区域识别，它通过使用较为浅层、较为简单的卷积神经网络快速生成人脸候选窗口。

R-Net（Refine Network，精化网络）使用了一个更为复杂的卷积神经网络，对 P-Net 初步得到的识别结果进行进一步的选择和调整，舍去大部分的错误输入。之后 R-Net 又再一次进行人脸区域的识别和关键点的定位，最终输出较为可信的识别结果。

O-Net(Output Network,输出网络)与 R-Net 类似,但它比 R-Net 更为复杂。O-Net 使用一个复杂的网络对模型进行优化,并输出最终的预测结果。

总的来说,MTCNN 的整体思想就是先用简单的方法得到一个粗略的结果,然后再用复杂的方法从粗略的结果里找到最终的结果。很显然,这一做法的好处是,使用简单的方法排除了大量的"错误答案"后,再用复杂的方法在大大缩小的范围内寻找"正确答案",从而大大提升了问题求解的效率。

5.2.2 安装操作

本节所用代码文件结构如表 5-2 所示。

表 5-2 代码文件结构

文 件 名 称	实 现 功 能
mtcnn.py	定义了神经网络的结构
main.py	实现使用现有模型进行人脸识别
utils.py	进行人脸识别需要用到的一些函数,包括检测人脸、生成人脸候选框

本次实战依赖的第三方库有 tensorflow、cv2、numpy、pillow 等。读者首先要在 Python 运行环境中安装好这些库,再在 Python 环境中运行 python main.py 对目标图像进行人脸检测。

5.2.3 代码解析

本次实战的核心代码包括 mtcnn.py、main.py 和 utils.py 3 个文件。在此主要针对 mtcnn.py 进行介绍,mtcnn.py 中实现了 P-Net、R-Net 和 O-Net 3 个网络的定义。

从代码中可以看出 P-Net 的网络结构是非常简单的,仅仅只有 3 层(此处没有计算用于得到人脸候选框以及得分的 2 个卷积层)。P-Net 依次经过 3 层神经网络,网络之间用使用 PReLU 激活函数,在第一层和第二层之间还有一个池化层。

```
class PNet(tf.keras.Model):
    def __init__(self):
        super().__init__()
        self.conv1 = tf.keras.layers.Conv2D(10, 3, 1, name='conv1')
        self.prelu1 = tf.keras.layers.PReLU(shared_axes=[1,2], name="PReLU1")
        self.conv2 = tf.keras.layers.Conv2D(16, 3, 1, name='conv2')
        self.prelu2 = tf.keras.layers.PReLU(shared_axes=[1,2], name="PReLU2")
        self.conv3 = tf.keras.layers.Conv2D(32, 3, 1, name='conv3')
        self.prelu3 = tf.keras.layers.PReLU(shared_axes=[1,2], name="PReLU3")
        self.conv4_1 = tf.keras.layers.Conv2D(2, 1, 1, name='conv4-1')
        self.conv4_2 = tf.keras.layers.Conv2D(4, 1, 1, name='conv4-2')
    def call(self, x, training=False):
        out = self.prelu1(self.conv1(x))
        out = tf.nn.max_pool2d(out, 2, 2, padding="SAME")
```

```python
        out = self.prelu2(self.conv2(out))
        out = self.prelu3(self.conv3(out))
        score = tf.nn.softmax(self.conv4_1(out), axis=-1)
        boxes = self.conv4_2(out)
        return boxes, score
```

R-Net 相比 P-Net 更为复杂,无论是网络层数(R-Net 有 4 层,而 P-Net 只有 3 层)还是每一层卷积网络的大小以及网络间的池化层和 flatten 层,R-Net 都比 P-Net 更为复杂。R-Net 处理来自 P-Net 的粗略结果,最后输出更为精确的预测结果交由 O-Net 处理。

```python
class RNet(tf.keras.Model):
    def __init__(self):
        super().__init__()
        self.conv1 = tf.keras.layers.Conv2D(28, 3, 1, name='conv1')
        self.prelu1 = tf.keras.layers.PReLU(shared_axes=[1,2], name="prelu1")
        self.conv2 = tf.keras.layers.Conv2D(48, 3, 1, name='conv2')
        self.prelu2 = tf.keras.layers.PReLU(shared_axes=[1,2], name="prelu2")
        self.conv3 = tf.keras.layers.Conv2D(64, 2, 1, name='conv3')
        self.prelu3 = tf.keras.layers.PReLU(shared_axes=[1,2], name="prelu3")
        self.dense4 = tf.keras.layers.Dense(128, name='conv4')
        self.prelu4 = tf.keras.layers.PReLU(shared_axes=None, name="prelu4")
        self.dense5_1 = tf.keras.layers.Dense(2, name="conv5-1")
        self.dense5_2 = tf.keras.layers.Dense(4, name="conv5-2")
        self.flatten = tf.keras.layers.Flatten()
    def call(self, x, training=False):
        out = self.prelu1(self.conv1(x))
        out = tf.nn.max_pool2d(out, 3, 2, padding="SAME")
        out = self.prelu2(self.conv2(out))
        out = tf.nn.max_pool2d(out, 3, 2, padding="VALID")
        out = self.prelu3(self.conv3(out))
        out = self.flatten(out)
        out = self.prelu4(self.dense4(out))
        score = tf.nn.softmax(self.dense5_1(out), -1)
        boxes = self.dense5_2(out)
        return boxes, score
```

O-Net 比 R-Net 更为复杂,它经过 5 层卷积神经网络,对来自 R-Net 的多个预测结果进行处理后得到最终的输出结果。

```python
class ONet(tf.keras.Model):
    def __init__(self):
        super().__init__()
        self.conv1 = tf.keras.layers.Conv2D(32, 3, 1, name="conv1")
        self.prelu1 = tf.keras.layers.PReLU(shared_axes=[1,2], name="prelu1")
        self.conv2 = tf.keras.layers.Conv2D(64, 3, 1, name="conv2")
```

```python
        self.prelu2 = tf.keras.layers.PReLU(shared_axes=[1,2], name="prelu2")
        self.conv3 = tf.keras.layers.Conv2D(64, 3, 1, name="conv3")
        self.prelu3 = tf.keras.layers.PReLU(shared_axes=[1,2], name="prelu3")
        self.conv4 = tf.keras.layers.Conv2D(128, 2, 1, name="conv4")
        self.prelu4 = tf.keras.layers.PReLU(shared_axes=[1,2], name="prelu4")
        self.dense5 = tf.keras.layers.Dense(256, name="conv5")
        self.prelu5 = tf.keras.layers.PReLU(shared_axes=None, name="prelu5")
        self.dense6_1 = tf.keras.layers.Dense(2 , name="conv6-1")
        self.dense6_2 = tf.keras.layers.Dense(4 , name="conv6-2")
        self.dense6_3 = tf.keras.layers.Dense(10 , name="conv6-3")
        self.flatten = tf.keras.layers.Flatten()
    def call(self, x, training=False):
        out = self.prelu1(self.conv1(x))
        out = tf.nn.max_pool2d(out, 3, 2, padding="SAME")
        out = self.prelu2(self.conv2(out))
        out = tf.nn.max_pool2d(out, 3, 2, padding="VALID")
        out = self.prelu3(self.conv3(out))
        out = tf.nn.max_pool2d(out, 2, 2, padding="SAME")
        out = self.prelu4(self.conv4(out))
        out = self.dense5(self.flatten(out))
        out = self.prelu5(out)
        score = tf.nn.softmax(self.dense6_1(out))
        boxes = self.dense6_2(out)
        lamks = self.dense6_3(out)
        return boxes, lamks, score
```

main.py 通过读取已有的网络模型和调用 utils.py 中实现的人脸检测方法对输入的图像进行人脸识别和特征点定位,在此不再赘述。

需要注意的是,本次实战的代码中并没有实现网络的训练过程,有兴趣的读者可以尝试自行实现。

5.2.4 训练测试

项目安装完成之后即可在命令行中输入 python main.py 对目标图像进行人脸检测。如果需要更换图像,可以在 main.py 中"image = cv2.cvtColor(cv2.imread("./multiface.jpg"), cv2.COLOR_BGR2RGB)"这行代码里将"./multiface.jpg"改为自己想要的图像的路径,如果想直接在命令行中指定要识别的图像路径或者对多张图像进行批量处理,可以尝试修改 main.py 文件自行实现。

输入 python main.py,代码运行的结果如图 5-3 所示,可以看到,MTCNN 可以准确地识别出一张图像中的各张人脸以及每张人脸上的关键点定位。

模型准确识别出了图像中的 6 张人脸,每张人脸以方框框出,同时在每张人脸上对 5 个关键点进行了定位,以点标出,分别是左眼、右眼、鼻尖、左嘴角和右嘴角。

图 5-3　运行命令与效果展示

第 6 章

单阶段目标检测

第 5 章中介绍了几个两阶段的目标检测的方法,本章中将对单阶段目标检测的几个经典方法进行实战。相比于两阶段目标检测的方法需要分两步进行区域框选和分类回归的任务,单阶段目标检测的方法会在一个阶段里完成这两项任务。因此,通常来说单阶段目标检测的方法在速度上优于两阶段目标检测的方法。本章将对 YOLOv4、SSD 和针对视频流的 DarkFlow 3 种单阶段目标检测方法进行实战。

6.1 应用单阶段完成目标检测之 YOLOv4

第 5 章介绍了 Faster RCNN 算法利用了两阶段结构,先实现了感兴趣区域的生成,再进行精细的分类与回归。Faster RCNN 虽出色地完成了物体检测任务,但两阶段的结构也限制了其速度,在更追求速度的实际应用场景下,应用起来仍不理想,在此背景下,YOLO 的 4 个版本应运而生。

6.1.1 背景原理

相比起 Faster RCNN 的两阶段结构,2015 年诞生的 YOLOv1 创造性地使用单阶段结构完成了物体检测任务,直接预测物体的类别与位置,没有 RPN 网络,也没有类似于 anchor 的预选框,因此速度很快。当然它也有不足之处:由于每一个区域默认只有两个边框做预测,并且只有一个类别,因此 YOLOv1 有着天然的检测限制。这种限制会导致对于小型物体,以及靠得特别近的物体检测效果不好。

针对 YOLOv1[26]的不足,2016 年诞生了 YOLOv2[27]。相比较 YOLOv1,YOLOv2 预测更加精准、速度更快,识别的物体类别也更多。2018 年又推出了 YOLOv3[28],在保持速度优势的前提下,进一步提升了检测精度,尤其是对小物体的检测能力。2020 年,Alexey 提出了 YOLOv4[29],YOLOv4 吸收了许多近年来方法的优势,在速度和检测质量上都有了很大提升,能够快速而准确地进行目标检测,并且硬件上只要有一块 1080Ti GPU 或者 2080Ti GPU 即可进行训练。得益于精准的预测效果以及高效的检测效果,YOLO 系列算法可以说是目前真正能够实现实时目标检测的最好算法。

除去输入层,YOLOv4 主要由 3 部分组成,即 BackBone、Neck 和 HEAD。

BackBone 部分是在数据集上进行预训练的骨干网络,YOLOv4 中针对图形处理单元(GPU)和视频处理单元(VPU)提供了两种神经网络:对于 GPU 使用少量组(1~8 组)的卷积,将 Darknet53 与 CSPNet 结合组成了 CSPDarknet53;而对于 VPU 则使用分组卷积。与 CSPNet 的结合提高了模型的学习能力,即使将模型轻量化也能保持预测的准确性,同时降低了对算力的要求和内存占用。

Neck 部分是为了扩大感受野以及更好地进行特征融合,使用了 SPP(空间金字塔池化)以及 PAN(路径汇聚网络)。

HEAD 部分则沿用了 YOLOv3 的 HEAD 部分进行稠密检测,输出预测内容和预测框。

6.1.2 安装操作

本节所用核心代码文件结构如表 6-1 所示。

表 6-1 代码文件结构

文 件 名 称	实 现 功 能	文 件 名 称	实 现 功 能
backbone.py	定义 CSPDarknet53 网络	benchmarks.py	与基准方法进行性能对比
common.py	公共数据准备实用工具	detect.py	实现图像目标检测
config.py	默认参数配置	detect_video.py	实现视频目标检测
dataset.py	定义数据集的类	evaluate.py	模型评估
utils.py	实用工具	save_model.py	转换模型
yolov4.py	YOLOv4 实现	train.py	模型训练

(1)安装依赖的第三方库:opencv、lxml、tqdm、tensorflow、absl、easydict、matplotlib、pillow 等,具体版本要求详见代码中的 requirements.txt 和 requirements-gpu.txt 两个文件。

(2)下载预训练好的权重文件,下载地址在 README.md 文件中给出,下载完成后放在 data 文件夹中。

(3)训练模型,在命令行中输入"python train.py --model yolov4 --weight 预训练权重地址",如果要使用轻量化的 YOLOv4,可以在命令中加上"--tiny"参数。

6.1.3 代码解析

本次实战的核心代码存放在 core 文件夹下,共有 backbone.py、common.py、config.py、dataset.py、utils.py 和 yolov4.py 6 个文件,以下针对 backbone.py、common.py 和 yolov4.py 3 个代码文件进行解析。

common.py 中实现了批规范化、卷积层结构、残差块、上采样等方法。YOLOv4 中用到的卷积层在 convolutional 函数中定义,代码如下。

```
def convolutional(input_layer, filters_shape, downsample=False, activate=True,
bn=True, activate_type='leaky'):
    #是否进行下采样
    if downsample:
        input_layer = tf.keras.layers.ZeroPadding2D(((1, 0), (1, 0)))(input_layer)
        padding = 'valid'
```

```python
            strides = 2
        else:
            strides = 1
            padding = 'same'
    #卷积层
    conv = tf.keras.layers.Conv2D(filters=filters_shape[-1],
            kernel_size = filters_shape[0], strides=strides,padding=padding, use
            _bias=not bn, kernel_regularizer=tf.keras.regularizers.l2(0.0005),
            kernel_initializer=tf.random_normal_initializer(stddev=0.01),
            bias_initializer=tf.constant_initializer(0.))(input_layer)
    #批正则化
    if bn: conv = BatchNormalization()(conv)
    #激活函数用 leaky 或 mish
    if activate == True:
        if activate_type == "leaky":
            conv = tf.nn.leaky_relu(conv, alpha=0.1)
        elif activate_type == "mish":
            conv = mish(conv)
    return conv
```

卷积层中用到的 mish 激活函数在 2019 年被首次提出，是当前最高水平的方法，其性能表现已经被验证超过了 ReLU，有可能打破 ReLU 统治深度学习激活函数的局面，mish 函数的代码实现如下。

```python
def mish(x):
    return x * tf.math.tanh(tf.math.softplus(x))
```

backbone.py 中实现了 darknet53、darknet53_tiny、cspdarknet53 以及 cspdarknet53_tiny 4 种进行预训练的骨干网络结构，在 YOLOv4.py 文件中实现了 YOLOv4 的全部网络。YOLOv4 的网络结构如图 6-1 所示，读者可以参照网络结构图理解 backbone.py 和 YOLOv4.py 两个文件中网络是如何组成的。

在 detect.py 文件中实现了对图像的目标检测，部分代码如下所示。

```python
#加载配置并以指定配置启动 TensorFlow 进程
config = ConfigProto()
config.gpu_options.allow_growth = True
session = InteractiveSession(config=config)
STRIDES, ANCHORS, NUM_CLASS, XYSCALE = utils.load_config(FLAGS)
input_size = FLAGS.size
image_path = FLAGS.image

#读入原图像并对图像进行归一化等处理
original_image = cv2.imread(image_path)
original_image = cv2.cvtColor(original_image, cv2.COLOR_BGR2RGB)
image_data = cv2.resize(original_image, (input_size, input_size))
```

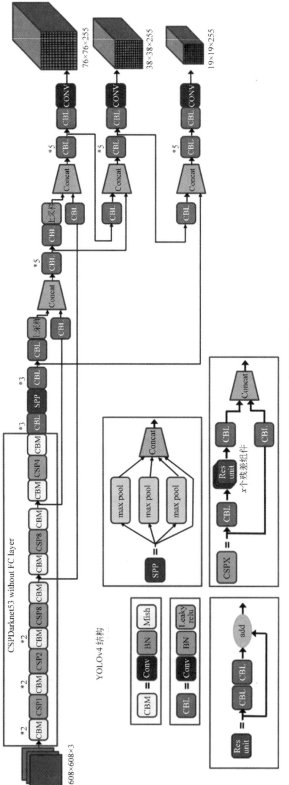

图 6-1 YOLOv4 的网络结构

```python
image_data = image_data / 255
images_data = []
for i in range(1):
    images_data.append(image_data)
images_data = np.asarray(images_data).astype(np.float32)

#加载保存的模型并进行目标检测
saved_model_loaded = tf.saved_model.load(FLAGS.weights, tags=[tag_constants.SERVING])
infer = saved_model_loaded.signatures['serving_default']
batch_data = tf.constant(images_data)
pred_bbox = infer(batch_data)
for key, value in pred_bbox.items():
    boxes = value[:, :, 0:4]
    pred_conf = value[:, :, 4:]
#对预测得到的预测结果进行重置等处理
boxes, scores, classes, valid_detections = tf.image.combined_non_max_suppression(
    boxes=tf.reshape(boxes, (tf.shape(boxes)[0], -1, 1, 4)),
    scores=tf.reshape(
        pred_conf, (tf.shape(pred_conf)[0], -1, tf.shape(pred_conf)[-1])),
    max_output_size_per_class=50,
    max_total_size=50,
    iou_threshold=FLAGS.iou,
    score_threshold=FLAGS.score
)
pred_bbox = [boxes.numpy(), scores.numpy(), classes.numpy(), valid_detections.numpy()]
#将检测结果画在原图像上显示出来并保存在本地
image = utils.draw_bbox(original_image, pred_bbox)
image = Image.fromarray(image.astype(np.uint8))
image.show()
image = cv2.cvtColor(np.array(image), cv2.COLOR_BGR2RGB)
cv2.imwrite(FLAGS.output, image)
```

6.1.4 训练测试

项目安装完成之后可以在命令行运行 detect.py 文件对图像进行目标检测。在运行 detect.py 文件之前，首先要对下载得到的预训练模型进行处理，在此使用项目中的 save_model.py 文件，处理预训练模型的语句是"python save_model.py --weights 预训练模型目录 -output 处理后的模型目录 -input_size 416 --model yolov4"，执行完成之后就得到了处理后的模型。如果需要使用轻量化的 YOLOv4，可以将"--model"后的参数改为 yolov4-tiny。

得到模型之后就可以用模型进行目标检测，输入"python detect.py --weights ./checkpoints/yolov4-416 --size 416 --model yolov4 --image ./data/kite.jpg"。这行命令的含

义是，使用"./checkpoint/yolov4-416"这个尺寸为416的YOLOv4模型对"./data/kite.jpg"文件进行目标检测，程序默认将结果图存放在当前目录下并命名为result.jpg，读者可以通过在命令中添加"— output 你指定的存放路径"来将结果保存在指定位置。上述命令运行后得到的目标检测结果如图6-2所示。

图6-2　YOLOv4目标检测结果

从图6-2中可以看出，YOLOv4方法将图像中的人、椅子、杯子、晚餐桌识别了出来，并且模型还给出了物体是某种目标的概率。从图6-2中可以看出，对于图像中的两张椅子，左边的椅子被以99%的概率识别为椅子，而右边的椅子仅被以66%的概率识别为椅子。可以直观地看到，左边的椅子被遮挡的面积比较小，而右边的椅子上坐了一个人，大量的面积都被人遮挡住了，这就不难理解，为何两把椅子被识别为椅子的概率相差如此之大了。除此之外，图像中还有窗户、挂钩、玩具娃娃等未被识别出来的物体，这是因为模型并没有针对这些目标进行训练。

6.2　应用锚定框完成目标检测之SSD

对于目标检测任务，前面讲解的Faster RCNN算法采用了两阶段检测框架：首先利用RPN网络进行感兴趣区域生成；然后再对该区域进行类别的分类与未知的回归，这种方法虽然显著提升了精度，但也限制了检测速度。YOLO算法利用回归的思想，使用单阶段网络络直接完成了目标检测，速度很快，但是精度有了明显的下降。

在此背景下，SSD（Single Shot Multibox Detector）[30]算法借鉴了Faster RCNN与YOLO的思想，在单阶段网络的基础上使用了固定框进行区域生成，并利用了多层的特征信息，在速度与检测精度上都有了一定的提升。作为单阶段网络，SSD算法从多个角度对目标检测做了创新，是一个简洁高效的检测网络。本节主要将利用SSD算法实现单阶段目标检测。

6.2.1　背景原理

SSD算法的思想，主要分为4方面。

（1）数据增强：SSD在数据增强部分做了充分的数据增强工作，包括光学变换与几何变换等，极大限度地扩充了数据集的丰富性，从而有效提升了模型的检测精度。

（2）网络骨架：SSD在原始VGGNet的基础上，进一步延伸了4个卷积模块，最深处的特征图大小为1×1，这些特征图具有不同的尺度与感受野，可以负责检测不同尺度的物体。

（3）PriorBox与多特征图：与Faster RCNN类似，SSD利用了固定大小与宽高的PriorBox作为生成区域，但与Faster RCNN不同的是，SSD不是只在一个特征图上设定预选框，而是在6个不同尺度上都设立预选框，并且在浅层特征图上设立较小的PriorBox来检测小物体，在深层特征图上设立较大的PriorBox来负责检测大物体。

（4）正、负样本的选取与损失计算：利用3×3的卷积在6个特征图上进行特征的提取，并分为分类与回归两个分支，代表所有预选框的预测试，随后进行预选框与真实框的匹配，利用IOU筛选出正样本与负样本，最终计算出分类损失与回归损失。SSD采用了VGG16作为基础网络，并在之上进行了一些改进，输入图像经过预处理后大小固定为300×300，首先经过VGG 16网络的前13个卷积层，然后利用两个卷积Conv 6与Conv 7取代原来的全连接网络，进一步提取特征。SSD算法的网络结构如图6-3所示。

6.2.2 安装操作

数据集：本书实现的代码版本支持Pascal VOC 2012数据集。Pascal VOC 2012数据集用于图像分类和目标检测两个任务的基准测试，主要包含20个常见类别，包含11 540张训练和验证图像，10 991张测试图像。为了对模型训练测试，需要以下步骤。

（1）下载Pascal VOC 2012数据集，网址如下。

http://host.robots.ox.ac.uk/pascal/VOC/

（2）将数据集加入项目中，目录结构如下。

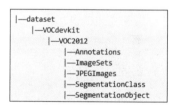

（3）安装依赖包：Python 3.6及以上版本、TensorFlow 2.1.0版本、numpy 1.17.0版本、opencv-python 4.1.0.25版本。

（4）运行write_voc_to_txt.py生成voct.txt。

```
python write_voc_to_txt.py
```

（5）运行train.py开始训练，在此之前，更改configuration.py中的参数值。

（6）在configuration.py中更改测试的单张图像目录test_picture_dir。

（7）运行test.py对单个图像进行测试。

6.2.3 代码解析

SSD是使用单个网络进行目标检测的统一框架。该项目的架构是模块化的，可以简化其他SSD变体（如基于ResNet或Inception）的实现和训练。目前的TF检查点已直接从SSD Caffe型号转换而来。该架构的灵感来自TF-Slim模型库，其中包含流行架构（ResNet、

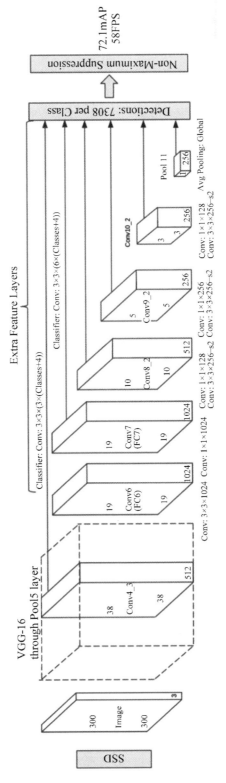

图 6-3　SSD 算法的网络结构

Inception 和 VGG)的实现。因此,它分为 3 个主要部分。DataSets:到流行数据集(Pascal VOC、COCO 等)的接口,用于将前者转换为 TF-Records 的脚本。网络:SSD 网络的定义,以及常用的编码和解码方法。预处理:预处理和数据增强例程。所以代码主要分为以下几种。

(1) 数据集准备:使用脚本将原始数据(如 VOC 数据集)转换为 tfrecord 文件。

(2) 数据预处理。目标:对输入数据进行数据增强,并转换到模型需要的形式。

注意点:对于图像分类任务,只需要对输入图像进行切片就行,标签不会发生变化;但对于物体检测任务来说,若对输入图像进行切片,那么输入的标签也会发生变化。这部分代码主要借鉴了 slim 的 preprocessing 模块。

(3) 模型搭建。目标:输入数据预处理结果,得到预测数据。SSD 模型前半部分几乎完全使用了图像分类模型(该源码中使用了 vgg 网络),后半部分从若干指定的 feature map 中通过 CNN 获取预测结果。

(4) 获取默认 anchor。目标:获取默认 anchor 信息。对于位置信息来说,模型预测的是 bbox 的偏移量,而非 bbox 本身。最终的 bbox 信息是通过默认 anchor 与 bbox 偏移量共同决定的。

(5) Ground Truth 预处理:为了计算损失函数,需要预测结果与 Ground Truth 一一对应。

(6) 损失函数:现在有了 Ground Truth 以及预测结果,就可以计算损失函数的值。这一模块不用于预测。

(7) post-processing(后处理)。目标:对于预测阶段,模型输出的结果比较多,需要筛选最终结果。

步骤对应的 Python 文件及方法,如表 6-2 所示。

表 6-2 步骤对应的 Python 文件及方法

处 理 步 骤	对应的 Python 文件	主 要 方 法
数据预处理	ssd_vgg_preprocessing.py	preprocess_for_train preprocess_for_eval
模型搭建	ssd_vgg_300.py、ssd_vgg_512.py	ssd_net、ssd_multibox_layer
获取默认 anchor	ssd_vgg_300.py、ssd_vgg_512.py	ssd_anchors_all_layers ssd_anchor_one_layer
Ground Truth 预处理	ssd_common.py	tf_ssd_bboxes_encode tf_ssd_bboxes_encode_layer
损失函数	ssd_vgg_300.py、ssd_vgg_512.py	ssd_losses
post-processing	np_methods.py	

1. 获取默认 anchor 实现细节

在源码中没有计算 anchor 面积的过程,而是直接指定了 anchor 的面积。对于长宽比为 1∶1 的 anchor 有特殊处理,这部分使用 numpy 实现。

1) sd_anchors_all_layers 解析

从源码中可知,具体实现是通过 ssd_anchor_one_layer 来完成的。

输入数据介绍如下。

（1）img_shape：表示输入数据的尺寸（经过 preprocessing 之后的结果）。

（2）layers_shape：要进行预测的特征图的尺寸列表。

（3）anchor_sizes：anchor 面积（本来是要进行计算，但这里直接给出数值），每个特征图分别获取。源码中，anchor 面积数量决定了 1∶1 的 anchor 数量。

（4）anchor_ratios：anchor 的长宽比例（不包括 1∶1 的 anchor），每个特征图分别获取。

（5）anchor_steps：特征图中每一点代表的实际像素值，每个特征图分别获取。

（6）offset：anchor 中心点在当前特征栅格中的位置。

每个特征图都有各自的 anchor_sizes、anchor_ratios、anchor_ratios、anchor_steps。

```
def ssd_anchors_all_layers(img_shape, layers_shape, anchor_sizes,anchor_ratios,
anchor_steps,offset=0.5,dtype=np.float32):
    layers_anchors = []
    for i, s in enumerate(layers_shape):
        anchor_bboxes = ssd_anchor_one_layer(img_shape, s, anchor_sizes[i],
anchor_ratios[i], anchor_steps[i],offset=offset, dtype=dtype)
        layers_anchors.append(anchor_bboxes)
    return layers_anchors
```

2）ssd_anchor_one_layer 解析

作用：获取指定特征图的所有默认 anchor。输出数据是 y、x、h、w。x、y 代表中心点位置，shape 为（N,M,1），其中（N,M）为特征图尺寸。h、w 代表边长，shape 为（N,），其中 N 为每个特征点的 anchor 数量。取值都为 0～1，都是在整张图像中的相对位置。

```
def ssd_anchor_one_layer(img_shape, feat_shape,sizes, ratios,step,offset=0.5,
                         dtype=np.float32):
    #计算 anchor 中心点位置
    y, x = np.mgrid[0:feat_shape[0], 0:feat_shape[1]]
    y = (y.astype(dtype) + offset) * step / img_shape[0]
    x = (x.astype(dtype) + offset) * step / img_shape[1]
    y = np.expand_dims(y, axis=-1)
    x = np.expand_dims(x, axis=-1)
    #计算相对边长
    #anchor 数量分为两部分
    #len(sizes) 代表长宽比为 1∶1 的 anchor 数量
    #len(ratios) 代表长宽比为 ratios 的 anchor 数量
    num_anchors = len(sizes) + len(ratios)
    h = np.zeros((num_anchors, ), dtype=dtype)
    w = np.zeros((num_anchors, ), dtype=dtype)
    #长宽比为 1∶1 的 anchor
    #从代码上看，1∶1 的 anchor 最多两个，不能更多了
    h[0] = sizes[0] / img_shape[0]
    w[0] = sizes[0] / img_shape[1]
    di = 1
    if len(sizes) > 1:
```

```
        h[1] = math.sqrt(sizes[0] * sizes[1]) / img_shape[0]
        w[1] = math.sqrt(sizes[0] * sizes[1]) / img_shape[1]
        di += 1
    #长宽比为 ratios 的 anchor
    for i, r in enumerate(ratios):
        h[i+di] = sizes[0] / img_shape[0] / math.sqrt(r)
        w[i+di] = sizes[0] / img_shape[1] * math.sqrt(r)
    return y, x, h, w
```

2. Ground Truth 预处理

1) tf_ssd_bboxes_encode 解析

功能的具体实现通过调用 tf_ssd_bboxes_encode_layer 完成。这部分主要介绍输入数据：labels 和 bboxes 是之前 preprocessing 的结果。anchor 是之前获取的所有默认 anchor，即一组 y、x、h、w 结果。

```
def tf_ssd_bboxes_encode(labels, bboxes, anchors, num_classes, no_annotation_
label, ig   nore_threshold=0.5, prior_scaling=[0.1, 0.1, 0.2, 0.2], dtype=tf.
float32,    scope='ssd_bboxes_encode'):
    with tf.name_scope(scope):
        target_labels = []
        target_localizations = []
        target_scores = []
        for i, anchors_layer in enumerate(anchors):
            with tf.name_scope('bboxes_encode_block_%i' % i):
                t_labels, t_loc, t_scores = \
                    tf_ssd_bboxes_encode_layer(labels, bboxes, anchors_layer,
                                    num_classes, no_annotation_label,
                                    ignore_threshold,
                                    prior_scaling, dtype)
                target_labels.append(t_labels)
                target_localizations.append(t_loc)
                target_scores.append(t_scores)
        return target_labels, target_localizations, target_scores
```

2) tf_ssd_bboxes_encode_layer 解析

```
def tf_ssd_bboxes_encode_layer(labels, bboxes, anchors_layer,num_classes,   no_
annotation_label,ignore_threshold=0.5,prior_scaling=[0.1, 0.1, 0.2, 0.2],
dtype=tf.float32):
    #`y, x, h, w`
    #`x, y`代表中心点位置,shape 为(N, M, 1),其中,(N, M)为特征图尺寸
    #`h, w`代表边长,shape 为(N, ),其中 N 为每个特征点的 anchor 数量
    #取值为 0~1,都是在整张图像中的相对位置
    yref, xref, href, wref = anchors_layer
    #转换 bbox 的表示方式
```

```python
        ymin = yref - href / 2.
        xmin = xref - wref / 2.
        ymax = yref + href / 2.
        xmax = xref + wref / 2.
        #计算 anchor 面积
        vol_anchors = (xmax - xmin) * (ymax - ymin)
        #shape 为(feature_map_height, feature_map_width, anchors_per_feature_map_
point)
        #可以代表特征图中所有 anchor
        shape = (yref.shape[0], yref.shape[1], href.size)
        feat_labels = tf.zeros(shape, dtype=tf.int64)
        feat_scores = tf.zeros(shape, dtype=dtype)
        feat_ymin = tf.zeros(shape, dtype=dtype)
        feat_xmin = tf.zeros(shape, dtype=dtype)
        feat_ymax = tf.ones(shape, dtype=dtype)
feat_xmax = tf.ones(shape, dtype=dtype)
#计算某个 bbox 与所有输入 anchor 的交并比
        def jaccard_with_anchors(bbox):
            int_ymin = tf.maximum(ymin, bbox[0])
            int_xmin = tf.maximum(xmin, bbox[1])
            int_ymax = tf.minimum(ymax, bbox[2])
            int_xmax = tf.minimum(xmax, bbox[3])
            h = tf.maximum(int_ymax - int_ymin, 0.)
            w = tf.maximum(int_xmax - int_xmin, 0.)
            #Volumes.
            inter_vol = h * w
            union_vol = vol_anchors - inter_vol \
                + (bbox[2] - bbox[0]) * (bbox[3] - bbox[1])
            jaccard = tf.div(inter_vol, union_vol)
            return jaccard
#计算某个 bbox 与 anchor 的交叉面积占 anchor 面积的比例
        def intersection_with_anchors(bbox):
        int_ymin = tf.maximum(ymin, bbox[0])
            int_xmin = tf.maximum(xmin, bbox[1])
            int_ymax = tf.minimum(ymax, bbox[2])
            int_xmax = tf.minimum(xmax, bbox[3])
            h = tf.maximum(int_ymax - int_ymin, 0.)
            w = tf.maximum(int_xmax - int_xmin, 0.)
            inter_vol = h * w
            scores = tf.div(inter_vol, vol_anchors)
            return scores
        def condition(i, feat_labels, feat_scores,
                      feat_ymin, feat_xmin, feat_ymax, feat_xmax):
            """ i < len(labels) """
            r = tf.less(i, tf.shape(labels))
```

```python
        return r[0]
    def body(i, feat_labels, feat_scores,
             feat_ymin, feat_xmin, feat_ymax, feat_xmax):
        label = labels[i]
        bbox = bboxes[i]
        jaccard = jaccard_with_anchors(bbox)
        #条件如下：cur_jaccard > scores && jaccard > jaccard_threshold && scores > -0.5 && label < num_classes
        mask = tf.greater(jaccard, feat_scores)
        mask = tf.logical_and(mask, feat_scores > -0.5)
        mask = tf.logical_and(mask, label < num_classes)
        imask = tf.cast(mask, tf.int64)
        fmask = tf.cast(mask, dtype)
#将符合条件的添加到 feat_labels/feat_scores/feat_ymin/feat_xmin/feat_ymax/feat_xmax 中，不符合条件的还是使用之前的值
        feat_labels = imask * label + (1 - imask) * feat_labels
        feat_scores = tf.where(mask, jaccard, feat_scores)
        feat_ymin = fmask * bbox[0] + (1 - fmask) * feat_ymin
        feat_xmin = fmask * bbox[1] + (1 - fmask) * feat_xmin
        feat_ymax = fmask * bbox[2] + (1 - fmask) * feat_ymax
        feat_xmax = fmask * bbox[3] + (1 - fmask) * feat_xmax
        return [i+1, feat_labels, feat_scores,
                feat_ymin, feat_xmin, feat_ymax, feat_xmax]
#本质就是遍历所有 Ground Truth 中的 label,将每个 gt 中的 label 与所有 anchor 进行对比。最后获取所有 anchor 的 label(分类标签)score(与 gt 的最大 jaccard)，以及 Groud Truth 的 bbox 信息
    i = 0
    [i, feat_labels, feat_scores,
     feat_ymin, feat_xmin,
     feat_ymax, feat_xmax] = tf.while_loop(condition, body,
                                            [i, feat_labels, feat_scores,
                                             feat_ymin, feat_xmin,
                                             feat_ymax, feat_xmax])
    #转换 bbox 表达方式
    feat_cy = (feat_ymax + feat_ymin) / 2.
    feat_cx = (feat_xmax + feat_xmin) / 2.
    feat_h = feat_ymax - feat_ymin
    feat_w = feat_xmax - feat_xmin
    #获取偏差值(预测数据就是预测偏差),并进行缩放
    feat_cy = (feat_cy - yref) / href / prior_scaling[0]
    feat_cx = (feat_cx - xref) / wref / prior_scaling[1]
    feat_h = tf.log(feat_h / href) / prior_scaling[2]
    feat_w = tf.log(feat_w / wref) / prior_scaling[3]
    feat_localizations = tf.stack([feat_cx, feat_cy, feat_w, feat_h], axis=-1)
    return feat_labels, feat_localizations, feat_scores
```

3. 损失函数

输入数据介绍如下。

（1）logits：anchor 分类预测结果，每个特征图一个数据。每个特征图中，数据 shape 为 [batch_size, feature_map_height, feature_map_width, num_anchors, num_classes]。

（2）localisations：anchor 的 bbox 预测结果，每个特征图一个数据。每个特征图中，shape 为 [batch_size, feature_map_height, feature_map_width, num_anchors, 4]。

（3）gclasses、glocalisations、gscores 均来自 Ground Truth 预处理。shape 均为 [feature_map_height, feature_map_width, num_anchors]。

通过指定 negative_ratio 实现正负样本的比例为 1∶3。

正反例分类误差分别计算。因为正例的标签是正数，反例的标签是 0。

```
def ssd_losses(logits, localisations,
               gclasses, glocalisations, gscores,
               match_threshold=0.5,
               negative_ratio=3.,
               alpha=1.,
               label_smoothing=0.,
               device='/cpu:0',
               scope=None):
    with tf.name_scope(scope, 'ssd_losses'):
        lshape = tfe.get_shape(logits[0], 5)
        num_classes = lshape[-1]
        batch_size = lshape[0]
        flogits = []
        fgclasses = []
        fgscores = []
        flocalisations = []
        fglocalisations = []
        for i in range(len(logits)):
            flogits.append(tf.reshape(logits[i], [-1, num_classes]))
            fgclasses.append(tf.reshape(gclasses[i], [-1]))
            fgscores.append(tf.reshape(gscores[i], [-1]))
            flocalisations.append(tf.reshape(localisations[i], [-1, 4]))
            fglocalisations.append(tf.reshape(glocalisations[i], [-1, 4]))
        logits = tf.concat(flogits, axis=0)
        gclasses = tf.concat(fgclasses, axis=0)
        gscores = tf.concat(fgscores, axis=0)
        localisations = tf.concat(flocalisations, axis=0)
        glocalisations = tf.concat(fglocalisations, axis=0)
        dtype = logits.dtype
        #根据gscores获取正/反例
        pmask = gscores > match_threshold
        fpmask = tf.cast(pmask, dtype)
```

```python
        n_positives = tf.reduce_sum(fpmask)
        #Hard negative mining...
        no_classes = tf.cast(pmask, tf.int32)
        predictions = slim.softmax(logits)
        nmask = tf.logical_and(tf.logical_not(pmask),
                               gscores > -0.5)
        fnmask = tf.cast(nmask, dtype)
        nvalues = tf.where(nmask,
                           predictions[:, 0],
                           1. - fnmask)
        nvalues_flat = tf.reshape(nvalues, [-1])
#设置反例数量为正例的 negative_ratio
        max_neg_entries = tf.cast(tf.reduce_sum(fnmask), tf.int32)
        n_neg = tf.cast(negative_ratio * n_positives, tf.int32) + batch_size
        n_neg = tf.minimum(n_neg, max_neg_entries)
        val, idxes = tf.nn.top_k(-nvalues_flat, k=n_neg)
        max_hard_pred = -val[-1]
        nmask = tf.logical_and(nmask, nvalues < max_hard_pred)
        fnmask = tf.cast(nmask, dtype)
#计算正例的分类误差
        with tf.name_scope('cross_entropy_pos'):
            loss = tf.nn.sparse_softmax_cross_entropy_with_logits(logits=logits,
                                                                  labels=gclasses)
            loss = tf.div(tf.reduce_sum(loss * fpmask), batch_size, name='value')
            tf.losses.add_loss(loss)
#计算反例的分类误差
        with tf.name_scope('cross_entropy_neg'):
            loss = tf.nn.sparse_softmax_cross_entropy_with_logits(logits=logits,
                                                                  labels=no_classes)
            loss = tf.div(tf.reduce_sum(loss * fnmask), batch_size, name='value')
            tf.losses.add_loss(loss)
        #bbox位置误差: smooth L1, L2,
        with tf.name_scope('localization'):
            weights = tf.expand_dims(alpha * fpmask, axis=-1)
            loss = custom_layers.abs_smooth(localisations - glocalisations)
            loss = tf.div(tf.reduce_sum(loss * weights), batch_size, name='value')
            tf.losses.add_loss(loss)
```

6.2.4 训练测试

为了训练出 SSD,首先要进行训练集的构建,再从数据集中读取数据并进行预处理。之后,通过模型获得位置信息与分类信息,并获取默认 anchor 信息,得到预测结果,接着对 Ground Truth 进行预处理(使得 Ground Truth 与预测结果一一对应)。最后,通过预测结果与 Ground Truth 计算损失函数,通过优化器进行训练。

预测流程:首先进行预测及构建,再从数据集中读取数据并进行预处理。通过模型获得位置信息与分类信息,并获取默认 anchor 信息,得到预测结果。将预测结果通过 post-processing(筛选 bbox 信息)获取最终预测结果(bbox 筛选后的结果)。

脚本 Train_SSD_network.py 负责训练网络。与 TF-Slim 模型类似,人们可以将许多选项传递给训练过程(数据集、优化器、超参数、模型等)。具体地说,可以提供可用作起始点以微调网络检查点文件。微调现有 SSD 检查点:优化 SSD 的最简单方法是将其用作预训练的 SSD 网络(VGG-300 或 VGG-512)。可以从前者开始对模型进行约束,如下所示。

```
DATASET_DIR=./tfrecords
TRAIN_DIR=./logs/
CHECKPOINT_PATH=./checkpoints/ssd_300_vgg.ckpt
python train_ssd_network.py \
    --train_dir=${TRAIN_DIR} \
    --dataset_dir=${DATASET_DIR} \
    --dataset_name=pascalvoc_2012 \
    --dataset_split_name=train \
    --model_name=ssd_300_vgg \
    --checkpoint_path=${CHECKPOINT_PATH} \
    --save_summaries_secs=60 \
    --save_interval_secs=600 \
    --weight_decay=0.0005 \
    --optimizer=adam \
    --learning_rate=0.001 \
    --batch_size=32
```

注意,除了训练脚本标志之外,读者可能还希望实验数据增强参数(随机裁剪、分辨率等)。这些都在 ssd_vgg_preprocessing.py 和网络参数(要素图层、锚定框等)中。

此外,训练脚本可以与评估例程相结合,以便监视验证数据集上保存的检查点的性能。为此,可以将 GPU 内存上限传递给训练和验证脚本,以便两者可以在同一设备上并行运行。如果有一些 GPU 内存可用于评估脚本,则可以按如下方式并行运行前者。

```
EVAL_DIR=${TRAIN_DIR}/eval
python eval_ssd_network.py \
    --eval_dir=${EVAL_DIR} \
    --dataset_dir=${DATASET_DIR} \
    --dataset_name=pascalvoc_2012 \
    --dataset_split_name=test \
    --model_name=ssd_300_vgg \
    --checkpoint_path=${TRAIN_DIR} \
    --wait_for_checkpoints=True \
    --batch_size=1 \
    --max_num_batches=500
```

微调接受过 ImageNet 训练的网络:还可以尝试基于标准架构(VGG、ResNet、Inception 等)构建新的 SSD 模型,并在其上面设置多个方块图层(具有特定的锚点、比率等)。为此,可以通过仅加载原始体系结构的权重来微调网络,并随机初始化网络的其余部分。例如,在 VGG-16 架构的情况下,可以按如下方式训练新模型。

```
DATASET_DIR=./tfrecords
TRAIN_DIR=./log/
CHECKPOINT_PATH=./checkpoints/vgg_16.ckpt
python train_ssd_network.py \
    --train_dir=${TRAIN_DIR} \
    --dataset_dir=${DATASET_DIR} \
    --dataset_name=pascalvoc_2012 \
    --dataset_split_name=train \
    --model_name=ssd_300_vgg \
    --checkpoint_path=${CHECKPOINT_PATH} \
    --checkpoint_model_scope=vgg_16 \

    --checkpoint_exclude_scopes=ssd_300_vgg/conv6,ssd_300_vgg/conv7,ssd_300_vgg/block8,ssd_300_vgg/block9,ssd_300_vgg/block10,ssd_300_vgg/block11,ssd_300_vgg/block4_box,ssd_300_vgg/block7_box,ssd_300_vgg/block8_box,ssd_300_vgg/block9_box,ssd_300_vgg/block10_box,ssd_300_vgg/block11_box \

    --trainable_scopes=ssd_300_vgg/conv6,ssd_300_vgg/conv7,ssd_300_vgg/block8,ssd_300_vgg/block9,ssd_300_vgg/block10,ssd_300_vgg/block11,ssd_300_vgg/block4_box,ssd_300_vgg/block7_box,ssd_300_vgg/block8_box,ssd_300_vgg/block9_box,ssd_300_vgg/block10_box,ssd_300_vgg/block11_box \
    --save_summaries_secs=60 \
    --save_interval_secs=600 \
    --weight_decay=0.0005 \
    --optimizer=adam \
    --learning_rate=0.001 \
    --learning_rate_decay_factor=0.94 \
    --batch_size=32
```

因此,在前面的命令中,训练脚本随机初始化属于 checkpoint_exclude_scope 的权重,并从检查点文件 vgg_16.ckpt 加载网络的其余部分。注意,在此还指定了 transable_scope 参数,首先只训练新的 SSD 组件,而保持 VGG 网络的其余部分不变。一旦网络收敛到较好的第一个结果,可以按如下方式微调整个网络。

```
DATASET_DIR=./tfrecords
TRAIN_DIR=./log_finetune/
CHECKPOINT_PATH=./log/model.ckpt-N
python train_ssd_network.py \
    --train_dir=${TRAIN_DIR} \
    --dataset_dir=${DATASET_DIR} \
    --dataset_name=pascalvoc_2012 \
    --dataset_split_name=train \
    --model_name=ssd_300_vgg \
    --checkpoint_path=${CHECKPOINT_PATH} \
```

```
--checkpoint_model_scope=vgg_16 \
--save_summaries_secs=60 \
--save_interval_secs=600 \
--weight_decay=0.0005 \
--optimizer=adam \
--learning_rate=0.00001 \
--learning_rate_decay_factor=0.94 \
--batch_size=32
```

在 TF-Slim 模型页面上可以找到一些流行的深层架构的预先训练的权重,页面地址为 https://github.com/tensorflow/models/tree/master/slim。

根据 Train_SSD_network.py 训练模型后,运行 test.py 得到部分测试结果如图 6-4 所示。

图 6-4 部分测试结果展示

图 6-4 中方框框出了检测目标的位置,框上方的单词表示了物体的分类。从图中可以看出来,SSD 的检测效果还是很不错的。同时做检测训练时还要注意:自己没有必要做简单的数据增强(如翻转等),SSD 训练过程中会做数据集增强训练的数据集。也不要用很模糊的目标,这样可能会导致训练发散,即 loss=NaN,因为 SSD 训练会根据 IOU 选取一定比率的样本作为正样本。综合大量实验结果来看,SSD 算法在准确度和速度上都比 YOLO 要好很多。YOLO 算法的缺点是难以检测小目标,而且定位不准。SSD 在 YOLO 的基础上主要改进了 3 点:多尺度特征图,利用卷积进行检测,设置先验框。这 3 点重要改进使得 SSD 在准确度上比 YOLO 更好,而且对于小目标检测效果也相对好一点。

6.3 基于视频流目标检测之 DarkFlow

在过去几年深度学习的发展背景下,图像目标检测任务取得了巨大的进展,检测性能得到明显提升。但在视频监控、车辆辅助驾驶等领域,基于视频的目标检测有着更为广泛的需求。本节主要介绍如何使用基于图像目标检测中的 YOLOv2 的 DarkFlow[31] 进行视频流目标检测。

6.3.1 背景原理

首先引入 YOLOv2,这是一个先进的可用于实时目标检测的模型。YOLOv2 使用了一个新的分类网络作为特征提取部分,参考了前人的先进经验,例如类似于 VGG,使用了较多的 3×3 卷积核,在每一次池化操作后把通道数翻倍。借鉴了 Network in Network 这篇论文中的 NIN 模型,网络使用了全局平均池化,把 1×1 的卷积核置于 3×3 的卷积核之间,用来压缩特征。也用了批量归一化来稳定模型训练。最终得出的基础模型就是 DarkNet-19,其包含 19 个卷积层、5 个最大值池化层。模型结构如图 6-5 所示。

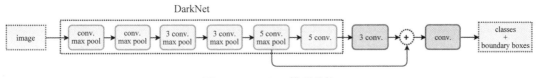

图 6-5　YOLOv2 模型结构

DarkNet-19 运算次数为 55.8 亿次,ImageNet 图像分类 top-1 准确率为 72.9%,top-5 准确率为 91.2%。要注意的是,YOLOv2 在 Pascal VOC 和 COCO 数据集上获得了很好的结果。采用多尺度训练方法,YOLOv2 可以根据速度和精确度需求调整输入尺寸。67FPS 时,YOLOv2 在 VOC2007 数据集上可以达到 76.8mAP;40FPS 时,可以达到 78.6mAP,比 Faster RCNN 和 SSD 精确度更高,检测速度更快。DarkFlow 实现了将 DarkNet 移植到 TensorFlow 上,可以用 TensorFlow 加载 DarkNet 训练好的模型,并使用 TensorFlow 重新训练,输出 TensorFlow Graph 模型,用于移动设备。接下来介绍如何使用 DarkFlow 进行视频流预测。

6.3.2 安装操作

本节所用的代码文件结构如表 6-3 所示。

表 6-3　代码文件结构

文件名称	实现功能	文件名称	实现功能
cfg	放置配置文件如 tiny-yolo.cfg 等	darkflow	实现 DarkFlow 模型
test	训练测试 DarkFlow	labels.txt	定义想要训练的标签

运行环境为 Python3,安装依赖库 numpy 和 opencv3。

(1) 可以选择以下 3 种方法之一开始使用 DarkFlow。只需要在适当的位置构建 Cython 扩展即可。注意:如果以这种方式安装,则必须使用克隆的 DarkFlow 目录中的 ./flow,而不是 flow,因为 DarkFlow 不是全局安装的。

```
python3 setup.py build_ext --inplace
```

(2) 让 pip 在 dev 模式下全局安装 DarkFlow(仍然可以全局访问,但对代码的更改会立即生效)。

```
pip install -e.
```

（3）使用 pip 全局安装。

```
pip install .
flow --h
```

首先，仔细看看一个非常有用的选项——load。

```
#1. Load tiny-yolo.weights
flow --model cfg/tiny-yolo.cfg --load bin/tiny-yolo.weights
#2. To completely initialize a model, leave the --load option
flow --model cfg/yolo-new.cfg
#3. It is useful to reuse the first identical layers of tiny for `yolo-new`
flow --model cfg/yolo-new.cfg --load bin/tiny-yolo.weights
#this will print out which layers are reused, which are initialized
```

来自默认文件夹 sample_img/ 的所有输入图像都流经网络，并将预测放入 sample_img/out 中。始终可以为此类正向传递指定更多参数，例如检测阈值、批次大小、图像文件夹等。

```
#Forward all images in sample_img/ using tiny yolo and 100% GPU usage
flow --imgdir sample_img/ --model cfg/tiny-yolo.cfg --load bin/tiny-yolo.weights --gpu 1.0
```

JSON 输出可以用每个边界框的像素位置和像素位置的描述来生成。默认情况下，每个预测都存储在 sample_img/out 文件夹中。下面展示了一个 JSON 数组示例。

```
#Forward all images in sample_img/ using tiny yolo and JSON output.
flow --imgdir sample_img/ --model cfg/tiny-yolo.cfg --load bin/tiny-yolo.weights --json
```

JSON 输出如下。

```
[{"label":"person", "confidence": 0.56, "topleft": {"x": 184, "y": 101},
"bottomright": {"x": 274, "y": 382}},
{"label": "dog", "confidence": 0.32, "topleft": {"x": 71, "y": 263},
"bottomright": {"x": 193, "y": 353}},
{"label": "horse", "confidence": 0.76, "topleft": {"x": 412, "y": 109},
"bottomright": {"x": 592,"y": 337}}]
```

6.3.3 代码解析

下面是 darknet.py 的代码解析。

```
from ..utils.process import cfg_yielder
from .darkop import create_darkop
from ..utils import loader
import warnings
import time
import os
```

```python
class Darknet(object):
    _EXT = '.weights'
    def __init__(self, FLAGS):
        self.get_weight_src(FLAGS)
        self.modify = False
        print('Parsing {}'.format(self.src_cfg))
        src_parsed = self.parse_cfg(self.src_cfg, FLAGS)
        self.src_meta, self.src_layers = src_parsed

        if self.src_cfg == FLAGS.model:
            self.meta, self.layers = src_parsed
        else:
            print('Parsing {}'.format(FLAGS.model))
            des_parsed = self.parse_cfg(FLAGS.model, FLAGS)
            self.meta, self.layers = des_parsed
        self.load_weights()
```

分析 FLAGS，加载以获取源二进制文件的位置及其配置。可以是 None、FLAGS、模型，或者其他。

```python
def get_weight_src(self, FLAGS):
    self.src_bin = FLAGS.model + self._EXT
    self.src_bin = FLAGS.binary + self.src_bin
    self.src_bin = os.path.abspath(self.src_bin)
    exist = os.path.isfile(self.src_bin)
    if FLAGS.load == str(): FLAGS.load = int()
    if type(FLAGS.load) is int:
        self.src_cfg = FLAGS.model
        if FLAGS.load: self.src_bin = None
        elif not exist: self.src_bin = None
    else:
        assert os.path.isfile(FLAGS.load), \
        '{} not found'.format(FLAGS.load)
        self.src_bin = FLAGS.load
        name = loader.model_name(FLAGS.load)
        cfg_path = os.path.join(FLAGS.config, name + '.cfg')
        if not os.path.isfile(cfg_path):
            warnings.warn(
                '{} not found, use {} instead'.format(
                cfg_path, FLAGS.model))
            cfg_path = FLAGS.model
        self.src_cfg = cfg_path
        FLAGS.load = int()
```

返回 layer 列表对象（darkop.py）给出 binaries/ 和 configs/ 的路径。

```python
def parse_cfg(self, model, FLAGS):
```

```
args = [model, FLAGS.binary]
cfg_layers = cfg_yielder(*args)
meta = dict(); layers = list()
for i, info in enumerate(cfg_layers):
    if i == 0: meta = info; continue
    else: new = create_darkop(*info)
    layers.append(new)
return meta, layers
```

使用 layer 和 loader 来加载权重文件。

```
def load_weights(self):
    print('Loading {} ...'.format(self.src_bin))
    start = time.time()
    args = [self.src_bin, self.src_layers]
    wgts_loader = loader.create_loader(*args)
    for layer in self.layers: layer.load(wgts_loader)
    stop = time.time()
    print('Finished in {}s'.format(stop - start))
```

6.3.4 训练测试

DarkFlow 的训练很简单,因为只需添加选项-train。如果这是第一次训练新配置,则将转换训练集和注释。要指向训练集和注释,可使用选项--DataSet 和--Annotation。下面是几个例子。

```
#从 yolo-tiny 中初始化 yolo-new,然后 100%在 GPU 上进行训练
flow --model cfg/yolo-new.cfg --load bin/tiny-yolo.weights --train --gpu 1.0
#完成 yolo-new 的初始化,使用 ADAM 进行训练
flow --model cfg/yolo-new.cfg --train --trainer adam
```

在训练期间,脚本偶尔会将中间结果保存到 TensorFlow 检查点,存储在 ckpt/中。要在执行训练/测试之前恢复到任何检查点,可使用--load[checkpoint_num]选项,如果checkpoint_num<0,则 DarkFlow 将通过解析 ckpt/checkpoint 加载最近的存储。

```
# Resume the most recent checkpoint for training
flow --train --model cfg/yolo-new.cfg --load -1
# Test with checkpoint at step 1500
flow --model cfg/yolo-new.cfg --load 1500
# Fine tuning yolo-tiny from the original one
flow --train --model cfg/tiny-yolo.cfg --load bin/tiny-yolo.weights
```

关于 Pascal VOC 2007 的训练示例如下。

```
# Download the Pascal VOC dataset:
curl -O https://pjreddie.com/media/files/VOCtest_06-Nov-2007.tar
tar xf VOCtest_06-Nov-2007.tar
# An example of the Pascal VOC annotation format:
```

```
vim VOCdevkit/VOC2007/Annotations/000001.xml
#Train the net on the Pascal dataset:
flow --model cfg/yolo-new.cfg --train --dataset "~/VOCdevkit/VOC2007/JPEGImages" --annotation "~/VOCdevkit/VOC2007/Annotations"
```

在自己的数据集上进行训练：下面的步骤假设要使用 Tiny YOLO，并且数据集有 3 个类，创建配置文件 Tiny-yolo-voc.cfg 的副本，并将其重命名为 Tiny-yolo-voc-3c.cfg。在 Tiny-yolo-voc-3c.cfg 中，将 Region 层（最后一层）中的类更改为要训练的类数。在示例中，设置类为 3。

```
[region]
anchors = 1.08,1.19,  3.42,4.41,  6.63,11.38,  9.42,5.11,  16.62,10.52
bias_match=1
classes=3
coords=4
num=5
softmax=1
```

在 mini-yolo-voc-3c.cfg 中，将卷积层（倒数第二层）中的滤波器更改为 num×(class+5)。在示例中，num 为 5，类为 3，5×(3+5)=40。因此，把滤波器数量设置为 40。

```
[convolutional]
size=1
stride=1
pad=1
filters=40
activation=linear
[region]
anchors = 1.08,1.19,  3.42,4.41,  6.63,11.38,  9.42,5.11,  16.62,10.52
```

更改 labels.txt 以包括想要训练的标签（标签的数量应该与在 mini-yolo-voc-3c.cfg 文件中设置的类别数量相同）。在示例中，labels.txt 将包含 3 个标签。

```
label1
label2
label3
```

当训练时，可参考 mini-yolo-voc-3c.cfg 模型。

```
flow --model cfg/tiny-yolo-voc-3c.cfg --load bin/tiny-yolo-voc.weights --train --annotation train/Annotations --dataset train/Images
```

为什么要保持原来的 mini-yolo-voc.cfg 文件不变呢？当 DarkFlow 看到正在加载 Tiny-yolo-voc.weights 时，它将在 cfg/文件夹中查找 Tiny-yolo-voc.cfg，并将该配置文件与设置的新配置文件--model cfg/iny-yolo-voc-3c.cfg 一一进行比较。在这种情况下，除最后两层外，每层的权重数量都完全相同。因此，它会将权重加载到直到最后两层的所有层中，因为它们现在包含不同数量的权重。

完全在 CPU 上运行的演示的摄像头/视频文件演示：

```
flow --model cfg/yolo-new.cfg --load bin/yolo-new.weights --demo videofile.avi
```

对于100%在GPU上运行的演示:

```
flow --model cfg/yolo-new.cfg --load bin/yolo-new.weights --demo videofile.avi --gpu 1.0
```

如果要使用网络摄像头/摄像头,只需要将Videofile.avi替换为关键字Camera即可。要保存带有预测边界框的视频,可添加--saveVideo选项。

使用来自另一个Python应用程序的DarkFlow时请注意,RETURN_RECAST(IMG)必须使用numpy.ndarray。图像必须预先加载并传递给RETURN_FORECT(IMG)。传递文件路径不起作用。RETURN_FORECT(IMG)的结果将是一个字典列表,以与上面列出的JSON输出相同的格式来表示每个检测到的对象的值。

```
from DarkFlow.net.build import TFNet
import cv2
options = {"model": "cfg/yolo.cfg", "load": "bin/yolo.weights", "threshold": 0.1}
tfnet = TFNet(options)
imgcv = cv2.imread("./sample_img/sample_dog.jpg")
result = tfnet.return_predict(imgcv)
print(result)
```

将构建的图形保存到protobuf文件(.pb)。

```
#保存最新的检查点到protobuf文件
flow --model cfg/yolo-new.cfg --load -1 --savepb
#保存图和权重到protobuf文件
flow --model cfg/yolo.cfg --load bin/yolo.weights --savepb
```

保存.pb文件时,还将生成一个.meta文件。这个.meta文件是元字典中所有内容的JSON转储,其中包含锚定框和标签等后处理信息。这样,从图表进行预测和进行后期处理所需的所有内容都包含在这两个文件中,而不需要使用.cfg或任何标签文件进行标记。创建的.pb文件可用于将图形迁移到移动设备(Java/C++/Objective-C++)。输入张量和输出张量的名称分别为input和output。要进一步了解这个协议文件的用法,可参考TensorFlow on C++ API的官方文档。要在iOS应用程序上运行它,只需要将该文件添加到Bundle Resources,并在源代码中更新该文件的路径。此外,DarkFlow支持从.pb和.meta文件加载以生成预测(而不是从.cfg和检查点或.weight加载)。

```
##Forward images in sample_img for predictions based on protobuf file
flow --pbLoad built_graph/yolo.pb --metaLoad built_graph/yolo.meta --imgdir sample_img/
```

如果希望在使用return_recast()时加载.pb和.meta文件,可以设置pbLoad和metaLoad选项来代替通常设置的model和load选项。

运行如下测试命令:

```
flow --model cfg/yolo-new.cfg --load bin/yolo-new.weights --demo videofile.avi
```

得到部分 avi 文件目标检测测试效果,如图 6-6 所示。

(a) 效果1

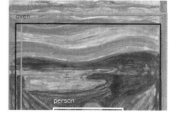
(b) 效果2

(c) 效果3

图 6-6　效果展示

效果图中用方框显示出了视频中某一帧时所有的目标位置,并给出了目标标签(如人)等。从图 6-6 中可以看出,这种视频目标检测方法能较好地保证在每帧图像上检测准确,并较好地保证了检测结果具有一致性和连续性,完整实现了视频流的人物目标识别。从模型结构来看,YOLO2 与 YOLO 相比主要有两方面的改进,在保持原有速度的同时提升了精度。它可以在各种图像大小下运行,且在各种检测数据集上比其他检测系统更快。但是也应该看到,由于视频中可能会存在运动模糊、遮挡、形态变化多样、光照变化多样等问题,所以在某些条件下利用图像目标检测技术检测视频中的目标并不能得到很好的检测结果。因此,如何利用视频中目标时序信息和上下文等信息成为未来提升视频目标检测性能的关键。

6.4　基于 ResNet 实现目标检测

在 RetinaNet[32] 之前,目标检测领域一个普遍的现象就是两阶段的方法有更高的准确率,但是耗时也更长,例如之前讲解的经典的 Faster RCNN 等。而单阶段的方法效率更高,但是准确性要差一些,例如之前讲解的经典的 YOLOv2、YOLOv3 和 SSD。这是两类方法本质上的思想不同导致的结果,而 RetinaNet 的出现,在一定程度上改善了这个问题,让单阶段的方法具备了比两阶段方法更高的准确性,而且耗时更低。因此,本节实现了基于

ResNet[33] 的 RetinaNet 模型目标检测。该模型通过修改 loss 函数处理了正负样本不均衡的问题，分析了单阶段算法和两阶段算法的差距，精度可以媲美两阶段算法。

6.4.1 背景原理

到目前为止，精度最高的物体检测器是基于 RCNN 推广的两阶段方法，其中分类器被应用于稀疏的候选物体位置集。相比之下，单阶段方法直接在原图上进行区域划分，暴力地进行分类和回归预测，这种简单粗暴的办法在速度上加快了不少，但也正是因为不够精细，在最后的结果表现上一直落后于两阶段方法。可以发现，在检测器的训练过程中遇到的正负样本不平衡是主要原因。因此，可以基于交叉熵损失函数，使用新的分类损失函数 Focal loss[34]，该损失函数通过抑制那些容易分类样本的权重，将注意力集中在那些难以区分的样本上，有效控制正负样本比例，防止失衡现象。为了验证 Focal loss 的有效性，已有的研究设计并训练了一个叫 RetinaNet 的网络进行评估设计。结果表明，RetinaNet 能够在实现和单阶段方法同等的速度基础上，在精度上超越所有（2017 年）双阶段的检测器。该模型采用 resnet50/resnet101 作为基础网络提取特征。之后用 FPN（特征空间金字塔）进行多尺寸的预测。共有 3 种尺寸的输出，每种输出为两路进行分类和 box 框的回归，输出时采用 9 个 anchor，结构如图 6-7 所示。

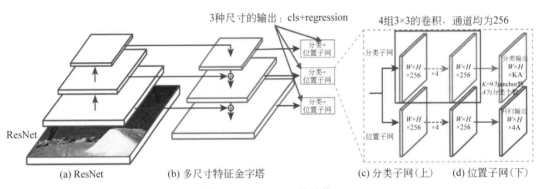

图 6-7　RetinaNet 的结构

6.4.2 安装操作

本节所用代码文件结构如表 6-4 所示。

表 6-4　代码文件结构

文 件 名 称	实 现 功 能
bin	实现模型转换、训练等
callbacks	定义召回率
models	实现 densenet、ResNet 等网络骨干
preprocessing	对数据进行预处理准备训练
utils	第三方库

续表

文 件 名 称	实 现 功 能
tests	测试所用数据与测试模型
ResNet50RetinaNet.ipynb	基于 ResNet50 骨干的测试示例

安装步骤如下。

(1) 进入 Anaconda Prompt 命令行,切换至本章根目录。

(2) 若要在 MS COCO 数据集上对模型进行训练或者测试时,需要先安装 pycotools,运行以下代码:

```
pip install --User
git+https://github.com/cocodataset/cocoapi.git#subd-irectory=PythonAPI;
```

(3) 在命令行中输入 jupyter notebook;

(4) 在浏览器中打开 ResNet50RetinaNet.ipynb;

分步运行代码,查看运行结果。

一般模型的工作过程如下。

```
Box,Score,Label=model.Forecast_On_Batch(输入)
```

其中,框 Box 大小为(None,None,4)(for(x1,y1,x2,y2)),分数 Score 大小为(None,None)(分类分数),标签 Label 大小为(None,None)(对应标签的分数)。3 个输出中第一个维度都表示形状,第二个维度索引检测列表。用以下命令加载模型。

```
from keras_retinanet.models import load_model
model = load_model('/path/to/model.h5', backbone_name='resnet50')
```

将训练模型转化为预测模型:在训练过程中使用训练模型。与预测模型相比,训练模型是比较精简的,只包含训练所需的层(回归和分类值)。如果希望对模型进行预测(对图像执行对象检测),则需要将训练好的模型转换为预测模型,运行以下命令。

```
#直接从库中运行
keras_retinanet/bin/convert_model.py /path/to/training/model.h5 /path/to/save/
inference/model.h5
#使用已安装脚本
retinanet-convert-model /path/to/training/model.h5 /path/to/save/inference/
model.h5
```

大多数脚本(如 retinanet-valuate)也支持使用--convert-model 参数进行动态转换。

使用 train.py 脚本对 keras-retinanet 进行训练。注意,因为这个训练脚本在 keras_retinanet 包中,所以使用相对导入。如果希望调整脚本以供自己在此代码库之外使用,则需要将其切换为绝对导入。如果正确安装了 keras-retinanet,则训练脚本将安装为 retinanet-train。但是,如果对 keras-retinanet 代码库进行本地修改,则应该直接从代码库运行脚本。这将确保训练脚本使用本地更改。默认骨干网络是 ResNet50。可以在运行的脚本中使用--bone=xxx 参数更改此设置。xxx 可以是 ResNet 模型(resnet50、resnet101、 resnet152)、Mobilenet[35]模型(Mobilenet128_1.0、Mobilenet128_0.75、Mobilenet160_1.0等)、Densenet模型或VGG模型中的骨干网络之一。不同的选项由每个模型在其相应的Python脚本(resnet.py、mobilenet.py等)中定义。训练好的模型不能直接用于预测，要将训练的模型转换为预测模型。

接着用不同数据集对模型进行训练：在Pascal VOC数据集上训练可以运行以下命令。

```
#直接从库中运行
keras_retinanet/bin/train.py pascal /path/to/VOCdevkit/VOC2007
#使用已安装脚本
retinanet-train pascal /path/to/VOCdevkit/VOC2007
```

在MS COCO数据集上训练可以运行以下命令。

```
#直接从库中运行
keras_retinanet/bin/train.py coco /path/to/MS/COCO
#使用已安装脚本
retinanet-train coco /path/to/MS/COCO
```

在OID数据集上训练可以运行以下命令。

```
#直接从库中运行
keras_retinanet/bin/train.py oid /path/to/OID
#使用已安装脚本
retinanet-train oid /path/to/OID
#如果想训练一个子集,也可以通过添加参数 labels_filter 来指定一个标签列表
keras_retinanet/bin/train.py oid /path/to/OID --labels-filter=Helmet,Tree
#如果希望在语义层次数的一个分支上进行训练,还可以通过指定父标签 labels_filter 实现
(https://storage.googleapis.com/openimages/challenge_2018/bbox_labels_500_
hierarchy_visualizer/circle.html)
keras_retinanet/bin/train.py oid /path/to/OID --parent-label=Boat
```

在KITTI上训练可以运行以下命令。

```
#直接从库中运行
keras_retinanet/bin/train.py kitti /path/to/KITTI
#使用已安装脚本
retinanet-train kitti /path/to/KITTI
```

如果想直接利用训练好的模型,可在以下地址中下载模型直接验证。

https://github.com/fizyr/keras-retinanet/releases

6.4.3 代码解析

本项目实战核心代码为训练代码和验证代码,分别为train.py、ResNet50RetinaNet.py。

1. train.py

使用create_models()函数创建3个模型,包括基本模型、训练模型和预测模型,其中backbone_retinanet表示用于创建具有给定骨干网络的retinanet模型函数。num_classes

表示训练种类，weights 表示模型加载权重，最终返回保存在 snapshot 中的基本模型、训练模型（如果参数 multi_gpu＝0，这个模型就跟基本模型相同）以及预测模型（包装执行物体检测的实用函数的模型，其中物体检测应用回归值并执行 NMS），具体实现过程如下。

```
def create_models(backbone_retinanet, num_classes, weights, multi_gpu=0,
                freeze_backbone=False, lr=1e-5, config=None):
    modifier = freeze_model if freeze_backbone else None
    #加载 anchor 参数，或者为 None 时使用默认参数
    anchor_params = None
    num_anchors   = None
    if config and 'anchor_parameters' in config:
        anchor_params = parse_anchor_parameters(config)
        num_anchors   = anchor_params.num_anchors()
#Keras 建议在 CPU 上初始化一个多 GPU 模型，以简化权重共享，并防止 OOM 错误。可以封装在并行模型中
if multi_gpu > 1:
        from keras.utils import multi_gpu_model
        with tf.device('/cpu:0'):
            model = model_with_weights(backbone_retinanet(num_classes, num_anchors=num_anchors, modifier=modifier), weights=weights, skip_mismatch=True)
        training_model = multi_gpu_model(model, gpus=multi_gpu)
    else:
        model=model_with_weights(backbone_retinanet(num_classes, num_anchors=num_anchors, modifier=modifier), weights=weights, skip_mismatch=True)
        training_model = model
    #创建预测模型
    prediction_model = retinanet_bbox(model=model, anchor_params=anchor_params)
    #编译模型
    training_model.compile(
        loss={
            'regression'    : losses.smooth_l1(),
            'classification': losses.focal()
        },
        optimizer=keras.optimizers.adam(lr=lr, clipnorm=0.001)
    )
    return model, training_model, prediction_model
```

在主函数中，首先，创建存储骨干信息的对象；其次，创建训练数据以及测试数据的 generator；最后，使用 model.fit_generator() 对模型进行训练。主要实现代码如下。

```
def main(args=None):
#创建存储 backbone 信息的对象
backbone = models.backbone(args.backbone)
    #创建训练数据以及测试数据的 generator
    train_generator, validation_generator = create_generators(args, backbone.
```

```python
        preprocess_image)
    #创建模型
    if args.snapshot is not None:
        print('Loading model, this may take a second...')
        model = models.load_model(args.snapshot, backbone_name=args.backbone)
        training_model    = model
        anchor_params     = None
        if args.config and 'anchor_parameters' in args.config:
            anchor_params = parse_anchor_parameters(args.config)
        prediction_model = retinanet_bbox(model=model, anchor_params=anchor_params)
    else:
        weights = args.weights
        #如果没有指定其他参数,则默认为 imagenet 权重
        if weights is None and args.imagenet_weights:
            weights = backbone.download_imagenet()
        print('Creating model, this may take a second...')
        model, training_model, prediction_model = create_models(
            backbone_retinanet=backbone.retinanet,
            num_classes=train_generator.num_classes(),
            weights=weights,
            multi_gpu=args.multi_gpu,
            freeze_backbone=args.freeze_backbone,
            lr=args.lr,
            config=args.config
        )
    #这让生成器可以使用实际的骨干网络模型来计算骨干网络层的形状
    if 'vgg' in args.backbone or 'densenet' in args.backbone:
        train_generator.compute_shapes = make_shapes_callback(model)
        if validation_generator:
            validation_generator.compute_shapes = train_generator.compute_shapes
    #在训练过程中使用 callback
    callbacks = create_callbacks(
        model,
        training_model,
        prediction_model,
        validation_generator,
        args,
    )
    if not args.compute_val_loss:
        validation_generator = None
    Return
training_model.fit_generator(generator=train_generator, steps_per_epoch=args.
ste    ps, epochs=args.epochs, verbose=1, callbacks=callbacks, workers=args.
```

```
workers, us e_multiprocessing=args.multiprocessing, max_queue_size=args.max_
queue_size,validation_data=validation_generator, initial_epoch=args.initial_
epoch)
```

2. ResNet50RetinaNet.py

加载 RetinaNet 模型,按照自己文件路径调整模型路径。

```
model = models.load_model(model_path, backbone_name='resnet50')
#如果模型没有转换为预测模型,那么可使用下面这一行
#model = models.convert_model(model)
#print(model.summary())
#为了可视化,将标签加载到名称映射
labels_to_names = {0: 'person', 1: 'bicycle', 2: 'car', 3: 'motorcycle', 4:
'airplane', 5: 'bus', 6: 'train', 7: 'truck', 8: 'boat', 9: 'traffic light', 10:
'fire hydrant', 11: 'stop sign', 12: 'parking meter', 13: 'bench', 14: 'bird', 15:
'cat', 16: 'dog', 17: 'horse', 18: 'sheep', 19: 'cow', 20: 'elephant'}
#在示例上运行检测
image = read_image_bgr('t3.jpg')
draw = image.copy()
draw = cv2.cvtColor(draw, cv2.COLOR_BGR2RGB)
#为网络预处理图像
image = preprocess_image(image)
image, scale = resize_image(image)
#处理时间
start = time.time()
boxes, scores, labels = model.predict_on_batch(np.expand_dims(image, axis=0))
print("processing time: ", time.time() - start)
#图像比例校正
boxes /= scale
#可视化检测
for box, score, label in zip(boxes[0], scores[0], labels[0]):
    #如果 scores 被分类就可以结束循环
    if score < 0.5:
        break
    color = label_color(label)
    b = box.astype(int)
    draw_box(draw, b, color=color)
    caption = "{} {:.3f}".format(labels_to_names[label], score)
    draw_caption(draw, b, caption)
plt.figure(figsize=(15, 15))
plt.axis('off')
plt.imshow(draw)
plt.show()
```

6.4.4 训练测试

运行 ResNet50RetinaNet.py 得到图 6-8 所示的多张目标检测效果图,框位置表明目标所处位置,数字表示目标分类正确的概率。

图 6-8 运行命令与效果展示

从图 6-8 中可以看出,模型可以实现以较大概率对目标进行正确分类。所以,总体来看,RetinaNet 就是一个基于 FPN 的单阶段检测器,靠着最后面的 Focal loss 来解决由于过量 background 而引起的类别不平衡问题。在图像目标预测中,正负样本比相差大,并且负样本大多为简单样本,大量的简单样本损失累计会导致训练缓慢,所以对预测错误的样本添加权重,从而使简单样本损失降得更大,从而优化了训练。从实验总体中看,RetinaNet 能够媲美先前的单阶段探测器的速度,同时超过现有的所有最先进的双阶段探测器的精度,在目标检测领域可以实现较好的应用。

第 7 章

实战文本识别和图像生成

前面已经介绍了应用深度神经网络的单阶段与两阶段的图像目标检测,现在来挑战将神经网络算法应用于文本,典型应用就是数字图像问题。本章介绍在 TensorFlow 2.0 环境下如何应用卷积神经网络实现识别手写体文本,利用自动编码器实现数字图像的压缩表示,并在此基础上实现变分自动编码器自动生成图像的实战。

7.1 应用卷积神经网络识别手写体文本

将 BP 网络用于图像处理的应用中,如果要对输入图像分类,可设计多层网络结构。如果采用全连接结构,存在参数爆炸问题。但如果利用局部卷积替换全连接结构,可明显减少参数,提高模型训练测试效率。本节主要利用三层的非线性神经网络模型对 MNIST 手写数据集中的数字进行识别,实现了较好的预测结果。

7.1.1 背景原理

将神经网络应用于文本中的一个典型应用,就是教会机器如何去自动识别图像中物体的种类。考虑图像分类中最简单的任务之一:0~9 数字图像识别,它相对简单,而且也具有非常广泛的应用价值,例如邮政编码数字识别等。而 Yann Lecun[36] 最早将 CNN 用于手写数字识别并一直保持了其在该问题的霸主地位。本节将以手写数字图像为例,探索使用卷积神经网络方法去解决这个问题。

如图 7-1 所示,收集大量的由真人书写的 0~9 的数字图像,为了便于存储和计算,一般把收集的原始图像缩放到某个固定的大小,比如 224 像素×224 像素,或者 96 像素×96 像素,这张图像将作为输入数据 x。同时,需要给每一张图像标注一个标签(Label)表明这张图像属于哪一个具体的类别。对于手写数字图像识别问题,编码更为直观,可以用数字的 0~9 来表示类别名字为 0~9 的图像。如果希望模型能够在新样本上也能具有良好的表现,即模型泛化能力较好,那么应该尽可能多地增加数据集的规模和多样性,使得用于学习的训练数据集与真实的手写数字图像的分布尽可能地逼近。

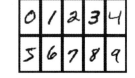

图 7-1 手写数字图像样例

本节采用了经业界统一测试和评估的手写数字图像数据集 MNIST,它包含了 0~9 共 10 种数字的手写图像,每种数字一共有 70 000 张图像,其中 60 000 张图像作为训练集训练模型,剩下 10 000 张图像作为测试集。考虑到手写数字图像包含的信息比较简单,每张图像均被缩放到 28 像素×28 像素的大小,同时只保留了灰度信息,如图 7-2 所示。

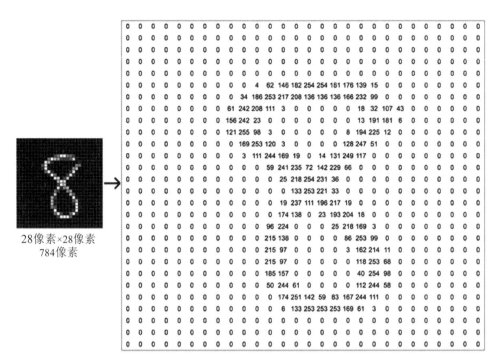

图 7-2　数字图像的表示

现在来看图像的表示方法,一张图像包含了 h 行,x 列,每个位置保存了像素值,像素值一般使用 0～255 的整形数值来表达颜色强度信息,例如 0 表示强度最低,255 表示强度最高。如果是彩色图像,则每个像素点包含了 R、G、B 3 个通道的强度信息,分别代表红色通道、绿色通道、蓝色通道的颜色强度,所以与灰度图像不同,它的每个像素点使用一个一维、长度为 3 的向量(Vector)来表示,向量的 3 个元素依次代表了当前像素点上面的 R、G、B 颜色强度值,因此彩色图像需要保存为形状是 $[h,x,3]$ 的张量。图 7-2 演示了内容为 8 的数字图像的矩阵内容,可以看到,图像中灰度越白的像素点,对应矩阵位置中数值也就越大。

7.1.2　安装操作

本节所用代码文件结构如表 7-1 所示。

表 7-1　代码文件结构

文 件 名 称	实 现 功 能
main.py	实现简单的卷积神经网络,完成 MNIST 数据集分类

运行 python main.py 命令。在这里用代码自动下载、管理和加载 MNIST 数据集,不需要额外编写代码,使用起来非常方便。

```
import  os
os.environ['TF_CPP_MIN_LOG_LEVEL']='2'
    #导入 TF 库
import  tensorflow as tf
```

```
from tensorflow import keras
from tensorflow.keras import layers, optimizers, datasets
    #加载数据集
(x, y), (x_val, y_val) = datasets.mnist.load_data()
```

7.1.3 代码解析

将下载的 MNIST 数据集转换为 NumPy 数组格式。

```
    #转换为张量,缩放到-1~1
x = tf.convert_to_tensor(x, dtype=tf.float32) / 255.
    #转换为张量
y = tf.convert_to_tensor(y, dtype=tf.int32)
    #one-hot 编码
y = tf.one_hot(y, depth=10)
print(x.shape, y.shape)
    #构建数据集对象
train_dataset = tf.data.Dataset.from_tensor_slices((x, y))
    #批量训练
train_dataset = train_dataset.batch(200)
```

load_data()函数返回两个元组(tuple)对象:第一个是训练集;第二个是测试集。每个元组的第一个元素是多个训练图像数据 X,第二个元素是训练图像对应的类别数字 Y。其中,训练集 X 的大小为(60000,28,28),代表了 60 000 个样本,每个样本由 28 行、28 列构成,由于是灰度图像,故没有 RGB 通道;训练集 Y 的大小为 60 000,代表了这 60 000 个样本的标签数字,每个样本标签用一个 0~9 的数字表示。测试集 X 的大小为(10000,28,28),代表了 10 000 张测试图像,Y 的大小为 10 000。

从 TensorFlow 中加载的 MNIST 数据图像,数值的范围为 0~255。在机器学习中,一般希望数据的范围在 0 周围小范围内分布。通过预处理步骤,把[0,255]像素范围归一化到[0,1]区间,再缩放到[-1,1]区间,从而有利于模型的训练。每张图像的计算流程是通用的,在计算的过程中可以一次进行多张图像的计算,充分利用 CPU 或 GPU 的并行计算能力。一张图像用 shape 为[h,w]的矩阵来表示,对于多张图像来说,在前面添加一个数量维度,使用 shape 为[c,h,x]。c 代表了 batch size。多张彩色图像可以使用 shape 为[c,h,x,d]的张量来表示,其中的 d 表示通道数量,彩色图像 $d=3$。通过 TensorFlow 的 Dataset 对象可以方便地完成模型的批量训练,只需要调用 batch()函数即可构建带 batch 功能的数据集对象。

网络搭建:在此采用三层的卷积神经网络对图像进行识别,例如对于第一层模型来说,接受的输入长度为 784 层,输出长度为 256 的向量,不需要显式地编写 ReLU 计算逻辑。使用 TensorFlow 的 Sequential 容器可以非常方便地搭建多层的网络。对于三层网络,可以通过

```
model = keras.Sequential([
    layers.Dense(512, activation='relu'),
    layers.Dense(256, activation='relu'),
```

```
layers.Dense(10)])
```

快速完成搭建,第一层的输出节点数设计为 256,第二层的输出节点数设计为 128,输出层节点数设计为 10。直接调用这个模型对象 model(x)就可以返回模型最后一层的输出。

7.1.4 训练测试

模型训练:得到模型输出后,通过 MSE 损失函数计算当前的误差 L。

```
with tf.GradientTape() as tape:
    #[b, 28, 28] => [b, 784]
    x = tf.reshape(x, (-1, 28 * 28))
    #第一步:计算输出
    #[b, 784] => [b, 10]
    out = model(x)
```

再利用 TensorFlow 提供的自动求导函数 tape.gradient(loss,model.trainable_variables)求出模型中所有的梯度信息

```
#第三步:优化,更新 w1, w2, w3, b1, b2, b3
grads = tape.gradient(loss, model.trainable_variables)
```

计算获得的梯度结果使用 grads 变量保存。再使用 optimizers 对象自动按着梯度更新法则去更新模型的参数。

```
optimizer.apply_gradients(zip(grads, model.trainable_variables))
```

循环迭代多次后,就可以利用学好的模型 g_θ 去预测未知的图像的类别概率分布。训练后得到手写数字图像 MNIST 数据集的训练误差曲线如图 7-3 所示。

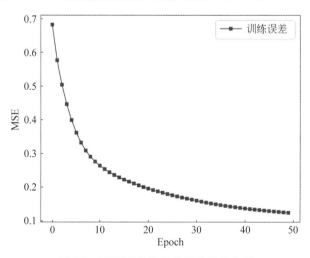

图 7-3　MNIST 数据集的训练误差曲线

由于三层的神经网络表达能力较强,手写数字图像识别任务简单,从图 7-3 中可以看到,误差值可以较快速、稳定地下降,其中对数据集的所有图像迭代一遍叫作一个 Epoch,可以在间隔数个 Epoch 后测试模型的准确率等指标,方便监控模型的训练效果。综上所述,

本节通过将一层的线性回归模型类推到分类问题,采用了表达能力更强的三层非线性神经网络,解决了手写数字图像识别的问题。

7.2 应用自动编码器识别手写体文本

在本节中,将使用自动编码器(Autoencoder)来对手写数字图像进行编码及解码。在此过程中,通过对样本的无监督学习,可以得到对手写数字图像的良好压缩表示,使用更少的空间来存储原有图像,并能够根据压缩表示,很好地进行图像重构。

7.2.1 背景原理

自动编码器(Autoencoder)[37][38]是一种无监督学习的神经网络算法。传统上,通常使用自动编码器来进行数据压缩和表示学习。自动编码器的网络结构如图 7-4 所示,输入 x 经过编码器(Encoder)进行编码,得到 x 的表示 z,z 是 x 的压缩表示,它的大小比 x 要小得多,是 x 的低维表示。再由解码器(Decoder)对 z 进行解码得到重构后的 x',通过比较 x' 与 x 之间的差异来不断进行优化,提高编码器和解码器的能力。

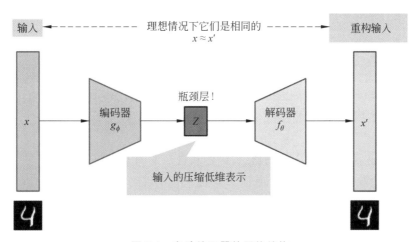

图 7-4 自动编码器的网络结构

一般在使用自动编码器时,通常使用它的前半部分,即编码器。使用编码器编码得到原来数据的良好压缩表示,并且这种表示能够通过解码器解码充分还原原始数据。自动编码器的最核心和精华的东西也就是这个表示,即图 7-4 中的 Z。

7.2.2 安装操作

本节所用代码文件结构如表 7-2 所示。

表 7-2 代码文件结构

文 件 名 称	实 现 功 能
main.py	实现自动编码器,完成 MNIST 数据集分类

安装步骤如下。

（1）安装 numpy、PIL、matplotlib 库。
（2）在 main.py 的路径下直接输入 python main.py 运行代码。

7.2.3 代码解析

本节的代码包括两部分内容。

第一部分：使用 MNIST 数据集，通过特征的压缩和解压，对比解压后的图像和压缩之前的图像是否一致，优化编码器和解码器，逐步实现两张图像越来越相似，在理想情况下两张图像最终应完全相等。

第二部分：输出编码器的结果，将输入输出后的图像进行可视化显示。

首先获取数据集，并进行数据预处理。

```
#获取MNIST手写数字数据集，并划分训练集和测试集
(x_train, y_train), (x_test, y_test) = keras.datasets.mnist.load_data()
#对图像进行归一化处理，将[0, 255]的像素取值缩放到[0, 1]
x_train, x_test = x_train.astype(np.float32) / 255., x_test.astype(np.float32) / 255.
```

接下来进行重要参数的设置。

```
#设置参数
image_size = 28 * 28         #图像尺寸，28像素×28像素
h_dim = 20                   #压缩的维数
num_epochs = 55              #训练轮数
batch_size = 100             #一个批次的大小
learning_rate = 1e-3         #学习率
```

定义自动编码器的类，实现编码器、解码器和自动编码器的调用。

```
#自动编码器类
class AE(tf.keras.Model):
    #初始化编码器和解码器的神经网络层
    def __init__(self):
        super(AE, self).__init__()
        #784 => 512
        self.fc1 = keras.layers.Dense(512)
        #512 => h
        self.fc2 = keras.layers.Dense(h_dim)
        #h => 512
        self.fc3 = keras.layers.Dense(512)
        #512 => image
        self.fc4 = keras.layers.Dense(image_size)
    #编码器将图像通过两个神经网络层从784维（28×28）压缩到512维，再压缩到h_dim维
    def encode(self, x):
        x = tf.nn.relu(self.fc1(x))
        x = (self.fc2(x))
        return x
```

```python
#解码器通过两个神经网络层对压缩后的图像进行恢复 h_dim=>512=>784
    def decode_logits(self, h):
        x = tf.nn.relu(self.fc3(h))
        x = self.fc4(x)
        return x
    #定义自动编码器的调用,进行图像的压缩与回复
    def call(self, inputs, training=None, mask=None):
        #encoder
        h = self.encode(inputs)
        #decode
        x_reconstructed_logits = self.decode_logits(h)
        return x_reconstructed_logits
#实现自动编码器,并用 Adam 优化器进行优化
model = AE()
model.build(input_shape=(4, image_size))
model.summary()
optimizer = keras.optimizers.Adam(learning_rate)


for epoch in range(num_epochs):
    for step, x in enumerate(dataset):
        x = tf.reshape(x, [-1, image_size])
        with tf.GradientTape() as tape:
            #调用一次自动编码器并计算重构损失
            x_reconstruction_logits = model(x)
            reconstruction_loss = tf.nn.sigmoid_cross_entropy_with_logits
(labels=x, logits=x_reconstruction_logits)
            reconstruction_loss = tf.reduce_sum(reconstruction_loss) / batch
_size
        #使用梯度对自动编码器进行优化
        gradients = tape.gradient(reconstruction_loss, model.trainable_variables)
        gradients, _ = tf.clip_by_global_norm(gradients, 15)
        optimizer.apply_gradients(zip(gradients, model.trainable_variables))
        #每 50 轮输出一次损失
        if (step + 1) % 50 == 0:
            print("Epoch[{}/{}], Step [{}/{}], Reconst Loss: {:.4f}"
                  .format(epoch + 1, num_epochs, step + 1, num_batches, float
(reconstruction_loss)))
        #每轮结束时保存重构的图像
    out_logits = model(x[:batch_size // 2])
    out = tf.nn.sigmoid(out_logits)#out is just the logits, use sigmoid
    out = tf.reshape(out, [-1, 28, 28]).numpy() * 255
    #将原图像与重构的图像进行拼接
    x = tf.reshape(x[:batch_size // 2], [-1, 28, 28])
    x_concat = tf.concat([x, out], axis=0).numpy() * 255.
    x_concat = x_concat.astype(np.uint8)
```

```
        index = 0
        for i in range(0, 280, 28):
            for j in range(0, 280, 28):
                im = x_concat[index]
                im = Image.fromarray(im, mode='L')
                new_im.paste(im, (i, j))
                index += 1
        #保存图像并显示
        new_im.save('images/vae_reconstructed_epoch_%d.png' % (epoch + 1))
        plt.imshow(np.asarray(new_im))
        plt.show()
        print('New images saved !')
```

7.2.4 训练测试

在命令行中输入 python main.py 后，程序运行得到图像经过编码和解码之后的对比效果如图 7-5 所示。

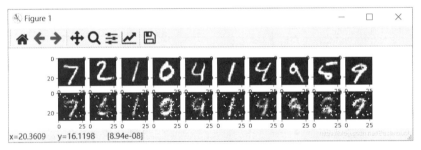

图 7-5 测试效果图

图 7-5 显示了手写数字图像经过编码和解码后得到的图像与原图像的对比效果，原图像在上，重构图像在下。从图中可以看出，重构后的图像与原图像是十分相似的，然而编码和解码的过程中仍然对图像质量造成了损失，重构后的图像相比原图像更加模糊，同时出现了很多噪点。

7.3 应用变分自动编码器重建服饰

变分自动编码器[39]与自动编码器十分相像，其不同点在于变分自动编码器对编码器添加了约束，强迫编码器产生服从标准高斯分布的潜在变量。正是由于这点不同，变分自动编码器拥有了生成样本的能力，是一个生成式模型，将实现使用变分自动编码器来进行图像的生成。

7.3.1 背景原理

自动编码器（AE）可以将图像进行压缩表示，并通过解码将压缩后的向量还原为一张图像。尽管自动编码器能够习得图像的低维表示，然而它不具备生成能力，习得的分布并不能用于产生新的图像,变分自动编码器(VAE)的出现就解决了这一问题。

变分自动编码器的模型结构如图7-6所示。变分自动编码器的不同之处在于它强迫编码器服从一个标准的高斯分布。因此,变分自动编码器的输出包含一个均值向量和一个标准差向量。在生成新的图像时,只要从高斯分布中进行采样传给解码器进行重构就可以了。在实际中,还需要在生成图像的精度和高斯分布的拟合度上进行权衡。

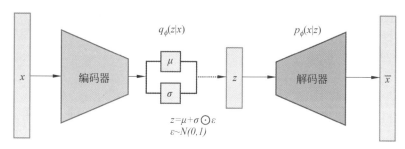

图 7-6 变分自动编码器的模型结构

从概率的角度看,变分自动编码器假设任何数据集都采样自某个分布 $p(x|z)$,z 是隐藏的变量,z 代表了某种内部特征,如手写数字的图像 x,z 可以表示字体的大小、书写风格、加粗、斜体等设定,它符合某个先验分布 $p(x|z)$,在给定具体隐藏变量 z 的情况下,可以从学到的分布 $p(x|z)$ 中采样一系列的生成样本,这些样本都具有 z 所表示的共性。在 $p(z)$ 已知(可以假定它符合已知的分布,如 $N(0,1)$)的条件下,目的就是希望能学会生成概率模型 $p(x|z)$。

本节所用的 Fashion MNIST 数据集即服饰数据集,由 70 000 张灰度图像组成,可以分成 T 恤、裤子等 10 个类别,类似于 MNIST 数据集,每个灰度图像都是 28 像素×28 像素的图像。

7.3.2 安装操作

本节所用代码文件结构如表 7-3 所示。

表 7-3 代码文件结构

文 件 名 称	实 现 功 能
main.py	实现变分自动编码器,完成服装重建

安装步骤如下。

(1) 安装 numpy、PIL、matplotlib 库。

(2) 在 main.py 的路径下直接输入 python main.py 运行代码。

7.3.3 代码解析

变分自动编码器和自动编码器的实现十分相似,本节将重点针对二者的不同之处进行代码解析。

首先是加载 Fashion MNIST 数据集以及设置参数,这部分和自动编码器是完全一样的。

```python
# 获取 Fashion MNIST 服饰数据集,并划分训练集和测试集
(x_train, y_train), (x_test, y_test) = keras.datasets.fashion_mnist.load_data()
# 对图像进行归一化处理,将[0, 255]的像素取值缩放到[0, 1]
x_train, x_test = x_train.astype(np.float32) / 255., x_test.astype(np.float32) / 255.
# 设置参数
image_size = 28 * 28          # 图像尺寸,28 像素×28 像素
h_dim = 20                    # 压缩的维数
num_epochs = 55               # 训练轮数
batch_size = 100              # 一个批次的大小
learning_rate = 1e-3          # 学习率
```

定义变分自动编码器的类,实现编码器、解码器和变分自动编码器的调用。

```python
# 变分自动编码器类
class VAE(tf.keras.Model):
    # 初始化编码器和解码器的神经网络层
    def __init__(self):
        super(VAE, self).__init__()
        # 编码器网络,与 AE 相比多了一个网络层,这是因为输出了均值和标准差两个向量
        self.fc1 = keras.layers.Dense(h_dim)
        self.fc2 = keras.layers.Dense(z_dim)
        self.fc3 = keras.layers.Dense(z_dim)
        # 解码器网络,与 AE 相同
        self.fc4 = keras.layers.Dense(h_dim)
        self.fc5 = keras.layers.Dense(image_size)
    # 编码器
    def encode(self, x):
        h = tf.nn.relu(self.fc1(x))
        return self.fc2(h), self.fc3(h)
    # 对编码器输出的均值和标准差向量进行处理,返回处理后的向量
    def reparameterize(self, mu, log_var):
        std = tf.exp(log_var * 0.5)
        eps = tf.random.normal(std.shape)
        return mu + eps * std
    # 解码器
    def decode_logits(self, z):
        h = tf.nn.relu(self.fc4(z))
        return self.fc5(h)
    def decode(self, z):
        return tf.nn.sigmoid(self.decode_logits(z))
    # 调用编码器和解码器
    def call(self, inputs, training=None, mask=None):
        # 编码
        mu, log_var = self.encode(inputs)
        # 采样
```

```python
            z = self.reparameterize(mu, log_var)
            #解码
            x_reconstructed_logits = self.decode_logits(z)
            return x_reconstructed_logits, mu, log_var
for epoch in range(num_epochs):
    for step, x in enumerate(dataset):
        x = tf.reshape(x, [-1, image_size])
        with tf.GradientTape() as tape:
            #前向传播
            x_reconstruction_logits, mu, log_var = model(x)
            #计算重构损失和KL散度
             reconstruction_loss = tf.nn.sigmoid_cross_entropy_with_logits(labels=x, logits=x_reconstruction_logits)
             reconstruction_loss = tf.reduce_sum(reconstruction_loss) / batch_size
            kl_div = - 0.5 * tf.reduce_sum(1. + log_var - tf.square(mu) - tf.exp(log_var), axis=-1)
            kl_div = tf.reduce_mean(kl_div)
            loss = tf.reduce_mean(reconstruction_loss) + kl_div
        #梯度优化
        gradients = tape.gradient(loss, model.trainable_variables)
        for g in gradients:
            tf.clip_by_norm(g, 15)
        optimizer.apply_gradients(zip(gradients, model.trainable_variables))
        #每50轮输出损失和KL散度
        if (step + 1) % 50 == 0:
            print("Epoch[{}/{}], Step [{}/{}], Reconst Loss: {:.4f}, KL Div: {:.4f}"
                .format(epoch + 1, num_epochs, step + 1, num_batches, float(reconstruction_loss), float(kl_div)))
    #生成模型
    z = tf.random.normal((batch_size, z_dim))
    out = model.decode(z)
    out = tf.reshape(out, [-1, 28, 28]).numpy() * 255
    out = out.astype(np.uint8)
    #保存图像
    index = 0
    for i in range(0, 280, 28):
        for j in range(0, 280, 28):
            im = out[index]
            im = Image.fromarray(im, mode='L')
            new_im.paste(im, (i, j))
            index += 1
    new_im.save('images/vae_sampled_epoch_%d.png' % (epoch + 1))
    plt.imshow(np.asarray(new_im))
    plt.show()
```

```
#保存重构的图像
out_logits, _, _ = model(x[:batch_size // 2])
out = tf.nn.sigmoid(out_logits)
out = tf.reshape(out, [-1, 28, 28]).numpy() * 255
x = tf.reshape(x[:batch_size // 2], [-1, 28, 28])
x_concat = tf.concat([x, out], axis=0).numpy() * 255.
x_concat = x_concat.astype(np.uint8)
index = 0
for i in range(0, 280, 28):
    for j in range(0, 280, 28):
        im = x_concat[index]
        im = Image.fromarray(im, mode='L')
        new_im.paste(im, (i, j))
        index += 1
new_im.save('images/vae_reconstructed_epoch_%d.png' % (epoch + 1))
plt.imshow(np.asarray(new_im))
plt.show()
print('New images saved !')
```

7.3.4 训练测试

在命令行输入 python main.py 后，程序运行会显示两组图像，如图 7-7 所示。

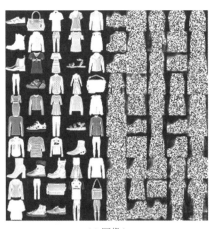

(a) 图像1　　　　　　　　　　　　(b) 图像2

图 7-7　训练测试效果图

图 7-7(a) 和图 7-7(b) 分别为在第 10、100 个 Epoch 时生成模型输出的图像和服装重构后的图像，每张图像的左 5 列为真实图像，右 5 列为对应的重建效果。从图中可以看出，VAE 作为生成模型，除了可以重建输入样本，还可以单独使用解码器生成样本。通过从先验分布 $p(z)$ 直接采样获取隐向量 z，经过解码后可以产生生成的样本，从而实现较为清晰的重建图像。

第 8 章

词汇与音乐的星空——机器创作

深度学习不仅能够在计算机视觉方面大放异彩,还可以进行音乐和歌词的创作。本章将更进一步讨论深度学习在自然语言和音乐方面的应用,主要介绍钢琴音乐的创作,具体包括了音乐生成和古诗创作两个主题,以期望能够通过实战案例让读者对基础的自然语言处理有一个初步的实践。

8.1 创作钢琴曲之 Music Transformer

创作音乐需要专业的领域知识和乐器技能,如何自动创作音乐已成为近年来一个热门研究方向。而将深度神经网络应用于音乐生成中,避免了传统机器学习中需要手工提取大量数据统计特征的时间耗费。因此,本节将利用 Music Transformer 基于注意的神经网络生成音乐序列,可直接创建"富有表现力"的表演而不需要先生成乐谱[40]。通过使用基于事件的表征和相对注意力的技术,Music Transformer 不仅能够更多地关注关系特征,而且能够超出训练样本的长度。由于它的内存密集程度较低,它还能生成更长的音乐序列。

8.1.1 背景原理

一首音乐作品通常由不同层次的反复出现的元素组成。因此,在很大程度上音乐利用重复来构建结构和意义。从主题旋律到具体音符等部分,要生成一个连贯的部分,模型需要参考已经出现的元素,重复、变化并进一步发展它们。

Transformer 是一种基于注意力的序列模型,在许多需要保持序列连贯性的生成任务中取得了很大成功。Transformer 本质上是一个 Encoder-Decoder 的结构,包括一个自注意力层以及一个前馈层。其中,注意力理论上可以建模任意长度的长距离依赖,并且符合人类直觉,具体模型结构如图 8-1 所示。Encoder 由 $N=6$ 个相同的 layer 组成,layer 指的就是图左侧的单元,最左边有个 Nx,这里是 6 个。图右侧的单元图是解码部分,同样地,Decoder 由 $N=6$ 个这样相同的层堆叠而成。

然而,在音乐创作和表演中,相对时序是很重要的。音乐有多个维度,沿着这些维度,相对差异和绝对差异相比显得更加有用,其中最突出的两个维度是时机和音高。为了捕捉表征之间的这种成对关系,需要使用两个位置之间的距离来调节自我注意力,这大大改进了样本质量。与原始的 Transformer 不同,具有相对注意力机制的 Transformer 样本保持了此数据集中存在的规则计时网格。此外,该模型还捕捉到了全局时序,从而产生了规则短语。

但是现有的这种基于成对距离来调整注意力的 Transformer,对于诸如音乐作品的长序列来说应用性也不够强,因为它们存储中间相对信息的复杂度是序列长度的平方。本次

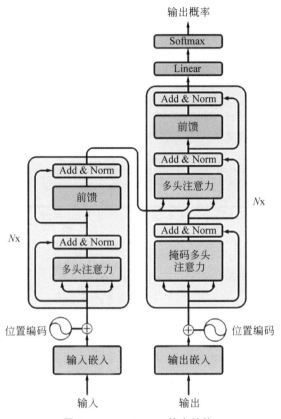

图 8-1 Transformer 基本结构

实战的算法实现了改进后的相对距离自注意力机制 Transformer，将它们的中间存储需求降低到序列长度量级，使得给定的 Transformer 能够在给定一个 motif 后生成数分钟长的音乐，或者基于 Seq2Seq 设置，给定一个旋律生成伴奏。

8.1.2 安装操作

本节所用代码文件结构如表 8-1 所示。

表 8-1 代码文件结构

文 件 名 称	实 现 功 能
train.py	加载数据并训练音乐模型
generate.py	运行模型并生成音乐

准备数据集：

```
python preprocess.py {midi_load_dir} {dataset_save_dir}
python train.py --epochs={NUM_EPOCHS} --load_path={NONE_OR_LOADING_DIR} --save_path={SAVING_DIR} --max_seq={SEQ_LENGTH} --pickle_dir={DATA_PATH} --batch_size={BATCH_SIZE} --l_r={LEARNING_RATE}
```

对图 8-2 的损失率曲线说明：最上面两条曲线为作为基准转换器的损失曲线；下面两条为音乐转换器的损失率曲线。从中可以看出随着轮次增加音乐转换器始终维持着较低的损失率。而与之对应着的图 8-3 则表明了音乐转换器相对于标准转换器有着更高的准确率。

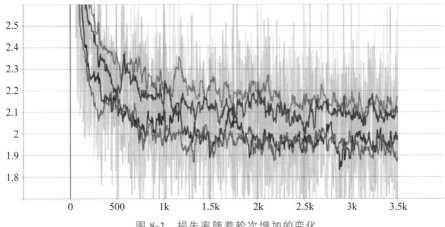

图 8-2　损失率随着轮次增加的变化

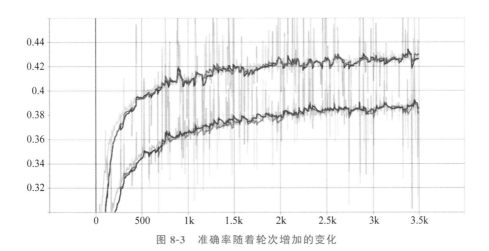

图 8-3　准确率随着轮次增加的变化

如下面代码所示，mt.Generate()可以自动生成音乐。

```python
from model import MusicTransformerDecoder
mt = MusicTransformerDecoder(
    embedding_dim=256, vocab_size=par.vocab_size,
    num_layer=6,
    max_seq=max_seq,
    dropout=0.1,
    debug=False
)
mt.generate(prior=[64], length=2048)
```

8.1.3 代码解析

本次实战的核心代码为 train.py 和 generate.py。首先分析训练过程代码 train.py。首先引入相关的库以及函数,并将参数初始化。

```
from model import MusicTransformerDecoder
from custom.layers import *
from custom import callback
import params as par
from tensorflow.python.keras.optimizer_v2.adam import Adam
from data import Data
import utils
import argparse
import datetime
import sys
tf.executing_eagerly()
parser = argparse.ArgumentParser()
parser.add_argument('--l_r', default=None, help='l_r', type=float)
parser.add_argument('--batch_size', default=2, help='batch size', type=int)
parser.add_argument('--pickle_dir', default='music', help='pickle_dir ')
parser.add_argument('--max_seq', default=2048, help='max_seq', type=int)
parser.add_argument('--epochs', default=100, help='epochs', type=int)
parser.add_argument('--load_path', default=None, help='load_path', type=str)
parser.add_argument('--save_path', default="result/dec0722", help=' save_path')
parser.add_argument('--is_reuse', default=False)
parser.add_argument('--multi_gpu', default=True)
parser.add_argument('--num_layers', default=6, type=int)
args = parser.parse_args()
```

然后进行参数设置,并加载数据和模型。

```
l_r = args.l_r
batch_size = args.batch_size
pickle_dir = args.pickle_dir
max_seq = args.max_seq
epochs = args.epochs
is_reuse = args.is_reuse
load_path = args.load_path
save_path = args.save_path
multi_gpu = args.multi_gpu
num_layer = args.num_layers
dataset = Data('dataset/processed')
print(dataset)
learning_rate = callback.CustomSchedule(par.embedding_dim) if l_r is None else l_r
opt = Adam(learning_rate, beta_1=0.9, beta_2=0.98, epsilon=1e-9)
```

接下来定义模型。

```python
mt = MusicTransformerDecoder(
        embedding_dim=256,
        vocab_size=par.vocab_size,
        num_layer=num_layer,
        max_seq=max_seq,
        dropout=0.2,
        debug=False, loader_path=load_path)
mt.compile(optimizer=opt, loss=callback.transformer_dist_train_loss)
```

定义可视化窗口。

```python
current_time = datetime.datetime.now().strftime('%Y%m%d-%H%M%S')
train_log_dir = 'logs/mt_decoder/'+current_time+'/train'
eval_log_dir = 'logs/mt_decoder/'+current_time+'/eval'
train_summary_writer = tf.summary.create_file_writer(train_log_dir)
eval_summary_writer = tf.summary.create_file_writer(eval_log_dir)
```

开始训练。

```python
idx = 0
for e in range(epochs):
    mt.reset_metrics()
    for b in range(len(dataset.files) // batch_size):
        try:
            batch_x, batch_y = dataset.slide_seq2seq_batch(batch_size, max_seq)
        except:
            continue
        result_metrics = mt.train_on_batch(batch_x, batch_y)
        if b % 100 == 0:
            eval_x, eval_y = dataset.slide_seq2seq_batch(batch_size, max_seq, 'eval')
            eval_result_metrics, weights = mt.evaluate(eval_x, eval_y)
            mt.save(save_path)
            with train_summary_writer.as_default():
                if b == 0:
                    tf.summary.histogram("target_analysis", batch_y, step=e)
                    tf.summary.histogram("source_analysis", batch_x, step=e)
                tf.summary.scalar('loss', result_metrics[0], step=idx)
                tf.summary.scalar('accuracy', result_metrics[1], step=idx)
            with eval_summary_writer.as_default():
                if b == 0:
                    mt.sanity_check(eval_x, eval_y, step=e)
                tf.summary.scalar('loss', eval_result_metrics[0], step=idx)
                tf.summary.scalar('accuracy', eval_result_metrics[1], step=idx)
                for i, weight in enumerate(weights):
                    with tf.name_scope("layer_%d" % i):
```

```
                    with tf.name_scope("w"):
                        utils.attention_image_summary(weight, step=idx)
```

当 with tf.name_scope("_w0")时,执行:

```
utils.attention_image_summary(weight[0])
```

当 with tf.name_scope("_w1")时,执行:

```
utils.attention_image_summary(weight[1])
            idx += 1
            print('\n==============================================')
            print('Epoch/Batch: {}/{}'.format(e, b))
            print('Train >>>> Loss: {:6.6}, Accuracy: {}'.format(result_metrics
[0], result_metrics[1]))
            print('Eval >>>> Loss: {:6.6}, Accuracy: {}'.format(eval_result_
metrics[0], eval_result_metrics[1]))
```

下面分析 generate.py。

```
from model import MusicTransformer, MusicTransformerDecoder
from custom.layers import *
from custom import callback
import params as par
from tensorflow.python.keras.optimizer_v2.adam import Adam
from data import Data
import utils
import datetime
import argparse
from midi_processor.processor import decode_midi, encode_midi
parser = argparse.ArgumentParser()
parser.add_argument('--max_seq', default=2048, help=' max_seq ', type=int)
parser.add_argument('--load_path', default="result/dec0722", help='load_path',
type=str)
parser.add_argument('--mode', default='dec')
parser.add_argument('--beam', default=None, type=int)
parser.add_argument('--length', default=2048, type=int)
parser.add_argument('--save_path', default='bin/generated.mid', type=str)
args = parser.parse_args()
```

参数设置如下所示。

```
max_seq = args.max_seq
load_path = args.load_path
mode = args.mode
beam = args.beam
length = args.length
save_path= args.save_path
```

```
current_time = datetime.datetime.now().strftime('%Y%m%d-%H%M%S')
gen_log_dir = 'logs/mt_decoder/generate_'+current_time+'/generate'
gen_summary_writer = tf.summary.create_file_writer(gen_log_dir)
```

训练模型：下面的 MusicTransformer 和 MusicTransformerDecoder 函数分别采用编解码网络和智能解码的方式进行模型的训练与保存工作。

```
if mode == 'enc-dec':
    print(">> generate with original seq2seq wise... beam size is {}".format(beam))
    mt = MusicTransformer(
            embedding_dim=256,
            vocab_size=par.vocab_size,
            num_layer=6,
            max_seq=2048,
            dropout=0.2,
            debug=False, loader_path=load_path)
else:
    print(">> generate with decoder wise... beam size is {}".format(beam))
    mt = MusicTransformerDecoder(loader_path=load_path)
inputs = encode_midi('dataset/midi/BENABD10.mid')
with gen_summary_writer.as_default():
    result = mt.generate(inputs[:10], beam=beam, length=length, tf_board=True)
    for i in result:
        print(i)
    if mode == 'enc-dec':
        decode_midi(list(inputs[-1 * par.max_seq:]) + list(result[1:]), file_path=save_path)
    else:
        decode_midi(result, file_path=save_path)
```

8.1.4 训练测试

音乐生成的命令如下。

```
python generate.py --load_path={CKPT_CONFIG_PATH} --length={GENERATE_SEQ_LENGTH} --beam={NONE_OR_BEAM_SIZE}
```

运行上述命令后，可以得到一段由神经网络训练出的模型创作出来的钢琴音乐，这段音乐保存在项目文件夹的根目录下，双击即可播放。

8.2 应用循环神经网络创作歌词

本节利用从零开始的循环神经网络搭建一个创作歌词的平台。

8.2.1 背景原理

音乐在生活中扮演着非常重要的作用，歌词的质量很大程度上决定了音乐的质量，因此歌词创作是非常重要的部分。近年来，随着神经网络的不断发展和完善，在各个领域都有着广泛的应用，特别是在自然语言处理方面。神经网络相比于传统的统计方法，能够更好地处理文本信息。因此，在本节采用循环神经网络的方法来创作歌词。

创作歌词最难的部分是使生成的歌词保持上下文意义的连贯性，传统的统计方法很难做到这一点，而在自然语言处理领域中，循环神经网络能很好地解决这个问题。

本节及后两节采用的数据集包含了周杰伦从第一张专辑 Jay 到第十张专辑《跨时代》中的歌词，并在后面几节里应用循环神经网络来训练语言模型。当模型训练好后，就可以用这个模型来创作歌词。

8.2.2 安装操作

本节所用代码文件结构如表 8-2 所示。

表 8-2 代码文件结构

文 件 名 称	实 现 功 能
rnn-scratch.ipynb	创作歌词的核心文件

本次实战的核心代码为 rnn-scratch.ipynb。

（1）进入 Anaconda Prompt 命令行。
（2）在本章根目录下输入命令 jupyter notebook。
（3）在浏览器中打开 rnn-scratch.ipynb。
（4）分步运行代码，即可得到程序运行结果。

8.2.3 代码解析

```
import tensorflow as tf
from tensorflow import keras
from tensorflow.keras import backend as f
import numpy as np
import sys
import time
sys.path.append("..")
import d2lzh_tensorflow2 as d2l
(corpus_indices, char_to_idx, idx_to_char, vocab_size) = d2l.load_data_jay_lyrics()
```

在 TensorFlow 中会自动调用 GPU 进行运算，故不需要指定 GPU。可以调用下面这个函数查看自己的计算机是否能够使用 GPU。

```
tf.test.gpu_device_name()
```

首先，为了将词表示成向量输入到神经网络，需要使用 one-hot 向量。假设词典中不同字符的数量为 N（即词典大小 vocab_size），每个字符已经同一个从 0 到 $N-1$ 的连续整数值索引一一对应。如果一个字符的索引是整数 i，那么创建一个全 0 的长为 N 的向量，并将位置为 i 的元素设成 1。该向量就是对原字符的 one-hot 向量。下面分别展示了索引为 0

和 2 的 one-hot 向量，向量长度等于词典大小。

```
tf.one_hot(np.array([0, 2]), vocab_size)
```

输出如下。

```
<tf.Tensor: id=4, shape=(2, 1027), dtype=float32, numpy=
array([[1., 0., 0., ..., 0., 0., 0.],
       [0., 0., 1., ..., 0., 0., 0.]], dtype=float32)>
```

每次采样的小批量的形状是（批量大小，时间步数）。下面的函数将这样的小批量变换成多个可以输入进网络的形状为（批量大小，词典大小）的矩阵，矩阵个数等于时间步数。也就是说，时间步 t 的输入为 $X_t \in \mathbb{R}^{n \times d}$，其中 n 为批量大小，d 为输入个数，即 one-hot 向量长度（词典大小）。

```
def to_onehot(X, size):              #本函数已保存在 d2lzh_tensorflow2 包中方便以后使用
    #X shape: (batch), output shape: (batch, n_class)
    return [tf.one_hot(x, size,dtype=tf.float32) for x in X.T]
X = np.arange(10).reshape((2, 5))
inputs = to_onehot(X, vocab_size)
len(inputs), inputs[0].shape
```

输出张量数量与大小为：

```
(5, TensorShape([2, 1027]))
inputs[0][1][5]
```

输出张量类型与 ID 为

```
<tf.Tensor: id=37, shape=(), dtype=float32, numpy=1.0>
```

接下来，初始化模型参数。隐藏单元个数 num_hiddens 是一个超参数。

```
num_inputs, num_hiddens, num_outputs = vocab_size, 256, vocab_size
def get_params():
    def _one(shape):
        return tf.Variable(tf.random.normal(shape=shape, stddev=0.01,mean=0,
dtype=tf.float32))
    #隐藏层参数
    W_xh = _one((num_inputs, num_hiddens))
    W_hh = _one((num_hiddens, num_hiddens))
    b_h = tf.Variable(tf.zeros(num_hiddens), dtype=tf.float32)
    #输出层参数
    W_hq = _one((num_hiddens, num_outputs))
    b_q = tf.Variable(tf.zeros(num_outputs), dtype=tf.float32)
    params = [W_xh, W_hh, b_h, W_hq, b_q]
    return params
```

根据循环神经网络的计算表达式实现该模型。首先定义 init_rnn_state 函数来返回初始化的隐藏状态。它返回由一个形状为（批量大小，隐藏单元个数）的值为 0 的 Array 组成

的元组。使用元组是为了更便于处理隐藏状态含有多个 Array 的情况。

```python
def init_rnn_state(batch_size, num_hiddens):
    return (tf.zeros(shape=(batch_size, num_hiddens)), )
```

下面的 rnn 函数定义了在一个时间步里如何计算隐藏状态和输出。这里的激活函数使用了 tanh 函数。当元素在实数域上均匀分布时，tanh 函数值的均值为 0。

```python
def rnn(inputs, state, params):
    #inputs 和 outputs 皆为 num_steps 个形状为(batch_size, vocab_size)的矩阵
    W_xh, W_hh, b_h, W_hq, b_q = params
    H, = state
    outputs = []
    for X in inputs:
        X=tf.reshape(X,[-1,W_xh.shape[0]])
        H = tf.tanh(tf.matmul(X, W_xh) + tf.matmul(H, W_hh) + b_h)
        Y = tf.matmul(H, W_hq) + b_q
        outputs.append(Y)
    return outputs, (H,)
state = init_rnn_state(X.shape[0], num_hiddens)
inputs = to_onehot(X, vocab_size)
params = get_params()
outputs, state_new = rnn(inputs, state, params)
print(len(outputs), outputs[0].shape, state_new[0].shape)
```

输出如下。

```
5 (2, 1027) (2, 256)
```

接下来定义预测函数。以下函数基于前缀 prefix（含有数个字符的字符串）来预测接下来的 num_chars 个字符。这个函数稍显复杂，其中将循环神经单元 RNN 设置成了函数参数，这样在后面小节介绍其他循环神经网络时能重复使用这个函数。

```python
#本函数已保存在 d2lzh_tensorflow2 包中方便以后使用
def predict_rnn(prefix, num_chars, rnn, params, init_rnn_state,
                num_hiddens, vocab_size,idx_to_char, char_to_idx):
    state = init_rnn_state(1, num_hiddens)
    output = [char_to_idx[prefix[0]]]
    for t in range(num_chars + len(prefix) - 1):
        #将上一时间步的输出作为当前时间步的输入
        X = tf.convert_to_tensor(to_onehot(np.array([output[-1]]), vocab_size),
dtype=tf.float32)
        X = tf.reshape(X,[1,-1])
        #计算输出和更新隐藏状态
        (Y, state) = rnn(X, state, params)
        #下一个时间步的输入是 prefix 里的字符或者当前的最佳预测字符
        if t < len(prefix) - 1:
            output.append(char_to_idx[prefix[t + 1]])
```

```
        else:
            output.append(int(np.array(tf.argmax(Y[0],axis=1))))
    #print(output)
    #print([idx_to_char[i] for i in output])
    return ''.join([idx_to_char[i] for i in output])
```

先测试一下 predict_rnn 函数。将根据"分开"创作长度为 10 个字符(不考虑前缀长度)的一段歌词。因为模型参数为随机值,所以预测结果也是随机的。

```
print(predict_rnn('分开', 10, rnn, params, init_rnn_state, num_hiddens, vocab_size,
                  idx_to_char, char_to_idx))
print(predict_rnn('不分开', 10, rnn, params, init_rnn_state, num_hiddens, vocab_size,
                  idx_to_char, char_to_idx))
```

10 轮初步训练后的输出为:

分开词担瘦 a 没已其妥四编
不分开词担瘦 a 没已其妥四编

能看到这里的输出结果并不成熟,因为训练轮次过于稀少。

```
#本函数已保存在 d2lzh 包中方便以后使用
def grad_clipping(grads,theta):
    norm = np.array([0])
    for i in range(len(grads)):
        norm+=tf.math.reduce_sum(grads[i] ** 2)
    #print("norm",norm)
    norm = np.sqrt(norm).item()
    new_gradient=[]
    if norm > theta:
        for grad in grads:
            new_gradient.append(grad * theta / norm)
    else:
        for grad in grads:
            new_gradient.append(grad)
    #print("new_gradient",new_gradient)
    return new_gradient
```

通常使用困惑度(Perplexity)来评价语言模型的好坏。困惑度是对交叉熵损失函数做指数运算后得到的值。特别地,最佳情况下,模型总是把标签类别的概率预测为 1,此时困惑度为 1;最坏情况下,模型总是把标签类别的概率预测为 0,此时困惑度为正无穷。基线情况下,模型总是预测所有类别的概率都相同,此时困惑度为类别个数。显然,任何一个有效模型的困惑度必须小于类别个数。在本例中,困惑度必须小于词典大小 vocab_size。定义模型训练函数跟之前章节的模型训练函数相比,这里的模型训练函数有以下几点不同。

(1)使用困惑度评价模型。

(2)在迭代模型参数前裁剪梯度。
(3)对时序数据采用不同采样方法将导致隐藏状态初始化的不同。

另外,考虑到后面将介绍的其他循环神经网络,为了更通用,这里的函数实现更长一些。

```python
def sgd(params, lr, batch_size, new_gradient):
    for i in range(len(params)):
        params[i].assign_sub((lr * new_gradient[i] / batch_size))
import math
#本函数已保存在d2lzh包中方便以后使用
def train_and_predict_rnn(rnn, get_params, init_rnn_state, num_hiddens,
                    vocab_size, corpus_indices, idx_to_char,
                    char_to_idx, is_random_iter, num_epochs, num_steps,
                    lr, clipping_theta, batch_size, pred_period,
                    pred_len, prefixes):
    if is_random_iter:
        data_iter_fn = d2l.data_iter_random
    else:
        data_iter_fn = d2l.data_iter_consecutive
    params = get_params()
    #loss = tf.keras.losses.SparseCategoricalCrossentropy()
    optimizer = tf.keras.optimizers.SGD(learning_rate=lr)
    for epoch in range(num_epochs):
        if not is_random_iter:       #如使用相邻采样,在epoch开始时初始化隐藏状态
            state = init_rnn_state(batch_size, num_hiddens)
        l_sum, n, start = 0.0, 0, time.time()
        data_iter = data_iter_fn(corpus_indices, batch_size, num_steps)
        for X, Y in data_iter:
            if is_random_iter:       #如使用随机采样,在每个小批量更新前初始化隐藏状态
                state = init_rnn_state(batch_size, num_hiddens)
            #else:                    #否则需要使用detach函数从计算图分离隐藏状态
                #for s in state:
                    #s.detach()
            with tf.GradientTape(persistent=True) as tape:
                tape.watch(params)
                inputs = to_onehot(X, vocab_size)
                #outputs有num_steps个形状为(batch_size, vocab_size)的矩阵
                (outputs, state) = rnn(inputs, state, params)
                #拼接之后形状为(num_steps * batch_size, vocab_size)
                outputs = tf.concat(outputs, 0)
                #Y的形状是(batch_size, num_steps),转置后再变成长度为
                #batch * num_steps 的向量,这样跟输出的行一一对应
                y = Y.T.reshape((-1,))
                #print(Y,y)
                y=tf.convert_to_tensor(y,dtype=tf.float32)
                #使用交叉熵损失计算平均分类误差
```

```
                l = tf.reduce_mean(tf.losses.sparse_categorical_crossentropy(y,
outputs))
                #l = loss(y,outputs)
                #print("loss",np.array(l))
            grads = tape.gradient(l, params)
            grads=grad_clipping(grads, clipping_theta)    #裁剪梯度
            optimizer.apply_gradients(zip(grads, params))
            #sgd(params, lr, 1 , grads)          #因为误差已经取过均值,梯度不用再做平均
            l_sum += np.array(l).item() * len(y)
            n += len(y)
        if (epoch + 1) % pred_period == 0:
            print('epoch %d, perplexity %f, time %.2f sec' % (
                epoch + 1, math.exp(l_sum / n), time.time() - start))
            #print(params)
            for prefix in prefixes:
                print(prefix)
                print(' -', predict_rnn(
                    prefix, pred_len, rnn, params, init_rnn_state,
                    num_hiddens, vocab_size,  idx_to_char, char_to_idx))
```

训练模型并创作歌词：现在可以训练模型了。首先，设置模型超参数。根据前缀"分开"和"不分开"分别创作长度为 50 个字符（不考虑前缀长度）的一段歌词。每过 50 个迭代周期便根据当前训练的模型创作一段歌词。

```
num_epochs, num_steps, batch_size, lr, clipping_theta = 250, 35, 32, 1e2, 1e-2
pred_period, pred_len, prefixes = 50, 50, ['分开', '不分开']
```

8.2.4 训练测试

下面采用随机采样训练模型并创作歌词。训练命令如下.

```
train_and_predict_rnn(rnn, get_params, init_rnn_state, num_hiddens,
                vocab_size, corpus_indices, idx_to_char,
                char_to_idx, True, num_epochs, num_steps, lr,
                clipping_theta, batch_size, pred_period, pred_len,
                prefixes)
```

训练的结果如下。

```
epoch 50, perplexity 75.588358, time 2.07 sec
epoch 100, perplexity 18.045482, time 2.05 sec
epoch 150, perplexity 7143.064714, time 2.82 sec
epoch 200, perplexity 4994.964191, time 2.06 sec
epoch 250, perplexity 1011.454457, time 2.26 sec
```

接下来采用相邻采样训练模型。

```
train_and_predict_rnn(rnn, get_params, init_rnn_state, num_hiddens,
```

```
                    vocab_size, corpus_indices, idx_to_char,
                    char_to_idx, False, num_epochs, num_steps, lr,
                    clipping_theta, batch_size, pred_period, pred_len,
                    prefixes)
epoch 50, perplexity 108.133573, time 2.14 sec
epoch 100, perplexity 42.373849, time 1.94 sec
epoch 150, perplexity 316711.357151, time 2.27 sec
epoch 200, perplexity 493805.452756, time 560.44 sec
epoch 250, perplexity 1145083.153934, time 1.98 sec
```

从上述结果可以看出训练模型的每个轮次和相关的复杂度数据，并且每一个轮次都在最后标注了该轮次训练所需要的时间。由于困惑度较高，因此还需要训练很多轮次才能得到有意义的结果。

- 分开始乡相信命运 让我碰到你 漂亮的让我面红的可爱女人感谢地心引力 温柔的让我心疼的可爱女人 透明的让
- 不分开 不能承受我已无处可躲 我不能再想 我不 我不 我不能 我不要再想 我不要再

上述是经过 400 次迭代轮次之后得到的机器创作歌词（这就是最后结果，但不通）。从生成内容上来看，实战中使用的基于字符级循环神经网络的语言模型生成歌词的语法和语义虽然还无法完全达到人类创作水平，但已经初步实现了简单歌词的完整创作。需要注意的是，训练循环神经网络时，为了应对梯度爆炸，可以使用裁剪梯度的方法。

8.3 应用门控循环单元创作歌词

8.2 节使用基于字符级的门控循环单元创作歌词，但是当时间步数较大或者时间步数较小时，循环神经网络的梯度较容易出现衰减或爆炸。虽然裁剪梯度可以应对梯度爆炸，但无法解决梯度衰减的问题。通常由于这个原因，循环神经网络在实际中较难捕捉时间序列中时间步距离较大的依赖关系。因此，本节使用门控循环单元（Gated Recurrent Unit，GRU）来捕捉时间序列中时间步距离较大的依赖关系。它通过学习门来控制信息的流动，进一步解决歌词创作中网络训练的梯度衰减问题。

8.3.1 背景原理

门控循环单元是一种常用的门控循环神经网络。它引入了重置门（Reset Gate）和更新门（Update Gate）的概念，从而修改了循环神经网络中隐藏状态的计算方式。如图 8-4 所示，门控循环单元中的重置门和更新门的输入均为当前时间步输入 X_t 与上一时间步隐藏状态 H_{t-1}，输出由激活函数为 sigmoid 函数的全连接层计算得到。

具体来说，假设隐藏单元个数为 h，给定时间步 t 的小批量输入 $X_t \in R^{n \times d}$（样本数为 n，输入个数为 d）和上一时间步隐藏状态 $H_{t-1} \in R^{n \times h}$。重置门 $R_t \in R^{n \times h}$ 和更新门 $Z_t \in R^{n \times h}$ 计算如下：$R_t = \sigma(X_t W_{xr} + H_{t-1} W_{hr} + b_r)$，$Z_t = \sigma(X_t W_{xz} + H_{t-1} W_{hz} + b_z)$，其中 $W_{xr}, W_{xz} \in R^{d \times h}$ 和 $W_{hr}, W_{hz} \in R^{h \times h}$ 是权重参数，$b_r \in R^{1 \times h}$ 是偏差参数。sigmoid 函数可以将元素的值变换到 0～1。因此，重置门 R_t 和更新门 Z_t 中每个元素的值域都是 [0,1]。

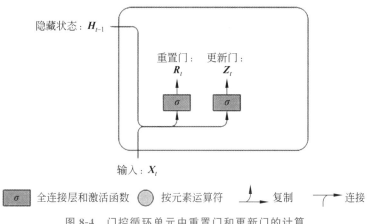

图 8-4 门控循环单元中重置门和更新门的计算

接下来,门控循环单元将计算候选隐藏状态来辅助稍后的隐藏状态计算。如图 8-5 所示,将当前时间步重置门的输出与上一时间步隐藏状态做按元素乘法(符号为 ⊙)。如果重置门中元素值接近 0,那么意味着重置对应隐藏状态元素为 0,即丢弃上一时间步的隐藏状态。如果元素值接近 1,那么表示保留上一时间步的隐藏状态。然后,将按元素乘法的结果与当前时间步的输入连接,再通过含激活函数 tanh 的全连接层计算出候选隐藏状态,其所有元素的值域为 [-1,1]。

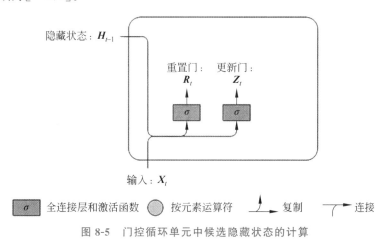

图 8-5 门控循环单元中候选隐藏状态的计算

具体来说,时间步 t 的候选隐藏状态 $H_t \in R^{h \times h}h$ 的计算为 $H_t = \tanh(X_t W_{xh} + (R_t \odot H_t - 1)W_{hh} + b_h)$,其中 $W_{xh} \in R^{d \times h}$ 和 $W_h \in R^{h \times h}$ 是权重参数,$b_h \in R^{1 \times h}$ 是偏差参数。从上面这个公式可以看出,重置门控制了上一时间步的隐藏状态如何流入当前时间步的候选隐藏状态。而上一时间步的隐藏状态可能包含了时间序列截至上一时间步的全部历史信息。因此,重置门可以用来丢弃与预测无关的历史信息。

最后,时间步 t 的隐藏状态 $H_t \in R^{n \times h}$ 的计算使用当前时间步的更新门 Z_t 来对上一时间步的隐藏状态 H_{t-1} 和当前时间步的候选隐藏状态 H_t 做组合。

值得注意的是,更新门可以控制隐藏状态应该如何被包含当前时间步信息的候选隐藏状态所更新,如图 8-6 所示。假设更新门在时间步 t' 到 $t(t' < t)$ 之间一直近似 1。那么,在

时间步 t' 到 t 之间的输入信息几乎没有流入时间步 t 的隐藏状态 H_t。实际上，这可以看作是较早时刻的隐藏状态 H_{t-1} 一直通过时间保存并传递至当前时间步 t。这个设计可以应对循环神经网络中的梯度衰减问题，并更好地捕捉时间序列中时间步距离较大的依赖关系。

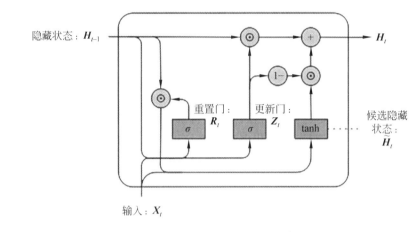

图 8-6 门控循环单元中隐藏状态的计算

在此对门控循环单元的设计进行总结。
（1）重置门有助于捕捉时间序列里短期的依赖关系。
（2）更新门有助于捕捉时间序列里长期的依赖关系。
为了实现并展示门控循环单元，本节仍使用周杰伦歌词数据集来训练模型作词。

8.3.2 安装操作

本节所用代码文件结构如表 8-3 所示。

表 8-3 代码文件结构

文 件 名 称	实 现 功 能
gru.ipynb	创作歌词的核心文件

本次实战的核心代码为 gru.ipynb。
（1）进入 Anaconda Prompt 命令行。
（2）本章根目录下输入命令 jupyter notebook。
（3）在浏览器中打开 gru.ipynb。
（4）分步运行代码，即可得到程序运行结果。

8.3.3 代码解析

首先，对模型参数进行初始化。超参数 num_hiddens 定义了隐藏单元的个数。

```
import tensorflow as tf
from tensorflow import keras
```

```python
import time
import math
import numpy as np
import sys
sys.path.append("..")
import d2lzh_tensorflow2 as d2l
(corpus_indices, char_to_idx, idx_to_char, vocab_size) = d2l.load_data_jay_
lyrics()
num_inputs, num_hiddens, num_outputs = vocab_size, 256, vocab_size
def get_params():
    def _one(shape):
        return tf.Variable(tf.random.normal(shape=shape,stddev=0.01,mean=0,
dtype=tf.float32))
    def _three():
        return (_one((num_inputs, num_hiddens)),
                _one((num_hiddens, num_hiddens)),
                tf.Variable(tf.zeros(num_hiddens), dtype=tf.float32))
    W_xz, W_hz, b_z = _three()    # 更新门参数
    W_xr, W_hr, b_r = _three()    # 重置门参数
    W_xh, W_hh, b_h = _three()    # 候选隐藏状态参数
    #输出层参数
    W_hq = _one((num_hiddens, num_outputs))
    b_q = tf.Variable(tf.zeros(num_outputs), dtype=tf.float32)
    #附上梯度
    params = [W_xz, W_hz, b_z, W_xr, W_hr, b_r, W_xh, W_hh, b_h, W_hq, b_q]
    return params
```

进一步定义隐藏状态初始化函数 init_gru_state。同 8.2 节中定义的 init_rnn_state 函数一样，它返回由一个形状为（批量大小，隐藏单元个数）的值为 0 的 NDArray 组成的元组。

```python
def init_gru_state(batch_size, num_hiddens):
    return (tf.zeros(shape=(batch_size, num_hiddens)), )
```

根据门控循环单元的计算表达式定义模型。

```python
def gru(inputs, state, params):
    W_xz, W_hz, b_z, W_xr, W_hr, b_r, W_xh, W_hh, b_h, W_hq, b_q = params
    H, = state
    outputs = []
    for X in inputs:
        X=tf.reshape(X,[-1,W_xh.shape[0]])
        Z = tf.sigmoid(tf.matmul(X, W_xz) + tf.matmul(H, W_hz) + b_z)
        R = tf.sigmoid(tf.matmul(X, W_xr) + tf.matmul(H, W_hr) + b_r)
        H_tilda = tf.tanh(tf.matmul(X, W_xh) + tf.matmul(R * H, W_hh) + b_h)
        H = Z * H + (1 - Z) * H_tilda
        Y = tf.matmul(H, W_hq) + b_q
```

```
        outputs.append(Y)
    return outputs, (H,)
```

接下来训练模型并创作歌词,在训练模型时只使用相邻采样。设置好超参数后,将训练模型并根据前缀"分开"和"不分开"分别创作长度为 50 个字符的一段歌词。

```
num_epochs, num_steps, batch_size, lr, clipping_theta = 160, 35, 32, 1e2, 1e-2
pred_period, pred_len, prefixes = 40, 50, ['分开', '不分开']
```

每过 40 个迭代周期便根据当前训练的模型创作一段歌词。

8.3.4 训练测试

```
d2l.train_and_predict_rnn(gru, get_params, init_gru_state, num_hiddens,
                          vocab_size, corpus_indices, idx_to_char,
                          char_to_idx, False, num_epochs, num_steps, lr,
                          clipping_theta, batch_size, pred_period, pred_len,
                          prefixes)
epoch 40, perplexity 585.433012, time 3.91 sec
epoch 80, perplexity 123.106788, time 4.11 sec
epoch 120, perplexity 52.382330, time 4.08 sec
epoch 160, perplexity 1046.588638, time 4.02 sec
```

简洁实现:在 Keras 中直接调用 layers 模块中的 GRU 类即可。

```
gru_layer = keras.layers.GRU(num_hiddens,time_major=True,return_sequences=True,return_state=True)
model = d2l.RNNModel(gru_layer, vocab_size)
d2l.train_and_predict_rnn_keras(model, num_hiddens, vocab_size,
                                corpus_indices, idx_to_char, char_to_idx,
                                num_epochs, num_steps, lr, clipping_theta,
                                batch_size, pred_period, pred_len, prefixes)
epoch 40, perplexity 141.880921, time 3.01 sec
epoch 80, perplexity 31.483826, time 2.88 se
epoch 120, perplexity 24.846631, time 2.94 sec
epoch 160, perplexity 165.971063, time 2.91 sec
```

下面是经过 160 次迭代轮次之后得到的机器创作歌词。

- 分开 还在祭榀蕃还我骷髅头像童话不达米亚平原几斤像养养 还在榀较翻还我骷髅头像童话不达米亚平原几斤像养养
- 不分开 一切限抄蜘蛛 还在榀杂草还我骷髅头像童话不达米亚平原几斤像养养 还在榀较翻还我骷髅头像童话不达米亚

从生成的歌词结果来看,由于门控循环单元引入了门的概念,从而修改了循环神经网络中隐藏状态的计算方式。其中包括重置门、更新门、候选隐藏状态和隐藏状态。重置门有助于捕捉时间序列里短期的依赖关系。更新门有助于捕捉时间序列里长期的依赖关系。因此,门控循环神经网络可以更好地捕捉时间序列中时间步距离较大的依赖关系,生成更好的歌词序列。

在上面的模型基础上我们可以调整一下参数,回答如下 3 个问题。

(1) 假设时间步 $t'<t$。如果只希望用时间步 t' 的输入来预测时间步 t 的输出,每个时间步的重置门和更新门的理想值是多少?

(2) 调节超参数,观察并分析对运行时间、困惑度以及创作歌词的结果造成的影响。

(3) 在相同条件下,比较门控循环单元和不带门控的循环神经网络的运行时间。

8.4 应用长短期记忆创作歌词

本节将介绍另一种常用的门控循环神经网络:长短期记忆(Long Short-Term Memory,LSTM)[41]。它比门控循环单元的结构稍微复杂一点,比 RNN 具备长期记忆功能,可控记忆能力。LSTM 是循环神经网络的变形结构,在普通 RNN 基础上,在隐藏层各神经单元中增加记忆单元,从而使时间序列上的记忆信息可控,每次在隐藏层各单元间传递时通过几个可控门(遗忘门、输入门、候选门、输出门),可以控制之前信息和当前信息的记忆和遗忘程度,从而使 RNN 网络具备了长期记忆功能,对于 RNN 的实际应用有巨大作用。

8.4.1 背景原理

LSTM 中引入了 3 个门,即输入门(Input Gate)、遗忘门(Forget Gate)和输出门(Output Gate),以及与隐藏状态形状相同的记忆细胞(某些文献把记忆细胞当成一种特殊的隐藏状态),从而记录额外的信息。

与门控循环单元中的重置门和更新门一样,如图 8-7 所示,长短期记忆的门的输入均为当前时间步输入 x_t 与上一时间步隐藏状态 h_{t-1},输出由激活函数为 sigmoid 函数的全连接层计算得到。如此一来,这 3 个门元素的值域均为[0,1]。

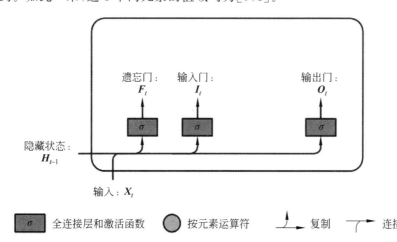

图 8-7 长短期记忆中输入门、遗忘门和输出门的计算

接下来,长短期记忆需要计算候选记忆细胞 C_t(见图 8-8)。它的计算与上面介绍的 3 个门类似,但使用了值域为[-1,1]的 tanh 函数作为激活函数。

如图 8-8 所示,可以通过元素值域在[0,1]的输入门、遗忘门和输出门来控制隐藏状态中信息的流动。这一般也是通过使用按元素乘法来实现。当前时间步记忆细胞的计算组合了上一时间步记忆细胞和当前时间步候选细胞的信息,并通过遗忘门和输入门来控制信息的流动。

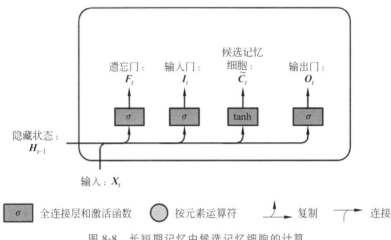

图 8-8　长短期记忆中候选记忆细胞的计算

如图 8-9 所示,遗忘门控制上一时间步的记忆细胞中的信息是否传递到当前时间步,而输入门则控制当前时间步的输入 X_t 候选记忆细胞。如果遗忘门一直近似 1 并按元素运算输入门一直近似 0,过去的记忆细胞将一直通过时间保存并传递至当前时间步。这个设计可以应对循环神经网络中的梯度衰减问题,并更好地捕捉时间序列中时间步距离较大的依赖关系。

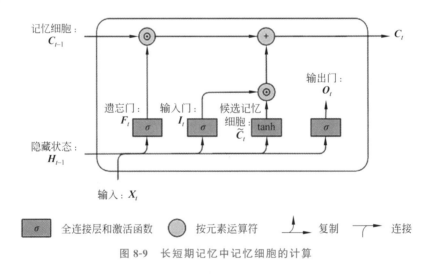

图 8-9　长短期记忆中记忆细胞的计算

有了记忆细胞后,接下来还可以通过输入门来控制从记忆细胞到隐藏状态 H_t 的信息流动。这里的 tanh 函数确保隐藏状态元素在 −1~1。需要注意的是,当输入门近似 1 时,记忆细胞信息将传递到隐藏状态供输出层使用;当输出门近似 0 时,记忆细胞信息只自己保留。图 8-10 展示了长短期记忆中隐藏状态的计算。

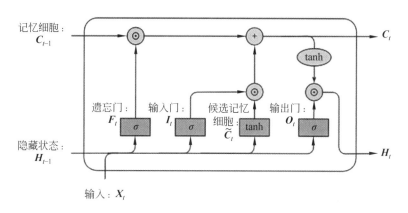

图 8-10 长短期记忆中隐藏状态的计算

8.4.2 安装操作

本节所用代码文件结构如表 8-4 所示。

表 8-4 代码文件结构

文 件 名 称	实 现 功 能
lstm.ipynb	创作歌词的核心文件

本次实战的核心代码为 lstm.ipynb。

（1）进入 Anaconda Prompt 命令行。

（2）在根目录下输入命令 jupyter notebook。

（3）在浏览器中打开 lstm.ipynb。

（4）分步运行代码，即可得到程序运行结果。

8.4.3 代码解析

首先，读取歌词数据集。

```
import tensorflow as tf
from tensorflow import keras
import time
import math
import numpy as np
import sys
sys.path.append("..")
import d2lzh_tensorflow2 as d2l
(corpus_indices, char_to_idx, idx_to_char, vocab_size) = d2l.load_data_jay_lyrics()
```

初始化模型参数。

```python
num_inputs, num_hiddens, num_outputs = vocab_size, 256, vocab_size
def get_params():
    def _one(shape):
        return tf.Variable(tf.random.normal(shape=shape, stddev=0.01, mean=0, dtype=tf.float32))
    def _three():
        return (_one((num_inputs, num_hiddens)),
                _one((num_hiddens, num_hiddens)),
                tf.Variable(tf.zeros(num_hiddens), dtype=tf.float32))
    W_xi, W_hi, b_i = _three()    #输入门参数
    W_xf, W_hf, b_f = _three()    #遗忘门参数
    W_xo, W_ho, b_o = _three()    #输出门参数
    W_xc, W_hc, b_c = _three()    #候选记忆细胞参数
    #输出层参数
    W_hq = _one((num_hiddens, num_outputs))
    b_q = tf.Variable(tf.zeros(num_outputs), dtype=tf.float32)
    return [W_xi, W_hi, b_i, W_xf, W_hf, b_f, W_xo, W_ho, b_o, W_xc, W_hc, b_c, W_hq, b_q]
```

定义模型。

```python
def init_lstm_state(batch_size, num_hiddens):
    return (tf.zeros(shape=(batch_size, num_hiddens)),
            tf.zeros(shape=(batch_size, num_hiddens)))
def lstm(inputs, state, params):
    W_xi, W_hi, b_i, W_xf, W_hf, b_f, W_xo, W_ho, b_o, W_xc, W_hc, b_c, W_hq, b_q = params
    (H, C) = state
    outputs = []
    for X in inputs:
        X=tf.reshape(X,[-1,W_xi.shape[0]])
        I = tf.sigmoid(tf.matmul(X, W_xi) + tf.matmul(H, W_hi) + b_i)
        F = tf.sigmoid(tf.matmul(X, W_xf) + tf.matmul(H, W_hf) + b_f)
        O = tf.sigmoid(tf.matmul(X, W_xo) + tf.matmul(H, W_ho) + b_o)
        C_tilda = tf.tanh(tf.matmul(X, W_xc) + tf.matmul(H, W_hc) + b_c)
        C = F * C + I * C_tilda
        H = O * tf.tanh(C)
        Y = tf.matmul(H, W_hq) + b_q
        outputs.append(Y)
    return outputs, (H, C)
```

8.4.4 训练测试

为了训练模型并创作歌词，实现以下代码。

```
num_epochs, num_steps, batch_size, lr, clipping_theta = 160, 35, 32, 1e2, 1e-2
```

```
pred_period, pred_len, prefixes = 40, 50, ['分开', '不分开']
d2l.train_and_predict_rnn(lstm, get_params, init_lstm_state, num_hiddens,
                          vocab_size, corpus_indices, idx_to_char,
                          char_to_idx, False, num_epochs, num_steps, lr,
                          clipping_theta, batch_size, pred_period, pred_len,
                          prefixes)
```

输出如下。

```
epoch 40, perplexity 3951.244398, time 4.65 sec
epoch 80, perplexity 1460.282746, time 4.68 sec
epoch 120, perplexity 526.141989, time 4.60 sec
epoch 160, perplexity 258.944899, time 4.57 sec
```

简单的实现后运行代码。

```
lr = 1e-2                              #注意调整学习率
lstm_layer = keras.layers.LSTM(num_hiddens,time_major=True,return_sequences=True,return_state=True)
model = d2l.RNNModel(lstm_layer, vocab_size)
d2l.train_and_predict_rnn_keras(model, num_hiddens, vocab_size, device,
                                corpus_indices, idx_to_char, char_to_idx,
                                num_epochs, num_steps, lr, clipping_theta,
                                batch_size, pred_period, pred_len, prefixes)
```

输出如下。

```
epoch 40, perplexity 1.020401, time 1.54 sec
epoch 80, perplexity 1.011164, time 1.34 sec
epoch 120, perplexity 1.025348, time 1.39 sec
epoch 160, perplexity 1.017492, time 1.42 sec
```

下面是经过 160 次迭代轮次之后得到的机器创作歌词。

- 分开始乡相信命运 感谢地心引力 让我碰到你 漂亮的让我面红的可爱女人 温柔的让我心疼的可爱女人 透明的让
- 不分开 我不能再想 我不 我不 我不能 爱情走的太快就像龙卷风 不能承受我已无处可躲 我不要再想 我不要再

8.5 应用文本生成器创作古诗

古诗生成是一个 RNN 的典型应用，也是 RNN 最能解决的一个应用场景。RNN 神经网络能够记住长序列中的某种特征，因此可以很好处理时序信息。RNN 可以处理多种时序信息，其中应用最广泛的是在文本上的处理，包含了文本情感分析、文本的自动生成。对于英文的诗歌的自动生成国外做得比较多，对于汉字的生成相对较少。中国古代的诗歌特别是唐诗宋词浩如烟海，唐诗宋词本身就有一定的内在规律，通过神经网络来发现这样的规律并表示出来就可以实现机器作诗。因此，本节以中国古诗为例，构造一个 RNN 模型，并用它来自动生成古诗。

8.5.1 背景原理

前几节已经介绍过循环神经网络,本节不再赘述。使用 RNN 得到古诗创作网络的输入是每一个汉字,总共有 1020 个字,用 one-hot 编码是一个 1020 维的稀疏向量。使用 one-hot 稀疏向量在输入层与网络第一层做矩阵乘法时会很没有效率,因为向量里面大部分都是 0,矩阵乘法浪费了大量的计算,最终矩阵运算得出的结果是向量中值为 1 的列所对应的矩阵中的行向量。

这看起来很像用索引查表一样,one-hot 向量中值为 1 的位置作为下标,去索引参数矩阵中的行向量。

为了代替矩阵乘法,将参数矩阵当作一个查找表(Lookup Table)或者叫作嵌入矩阵(Embedding Matrix),使用每个汉字所对应索引。如图 8-11,one-hot 向量中值为 1 的位置作为下标,去索引参数矩阵中的行向量。例如汉字"你"的索引值是 958,然后在查找表中找第 958 行即可。

图 8-11 预测网络

这其实跟替换之前的模型没有什么不同,嵌入矩阵就是参数矩阵,嵌入层仍然是隐层。查找表只是矩阵乘法的一种便捷方式,它会像参数矩阵一样被训练,是要学习的参数。

下面就是要构建的网络架构,从嵌入层输出的向量进入 LSTM 层进行时间序列的学习,然后经过 softmax 预测出下一个汉字。本节所用训练数据集为搜集到的 40 000 多首的唐诗,部分数据显示如图 8-12 所示。

图 8-12 古诗生成训练数据集示例

8.5.2 安装操作

本节所用代码文件结构如表 8-5 所示。

表 8-5 代码文件结构

文件名称	实现功能
poetry_generation_tf2.ipynb	创作古诗的核心文件

本次实战的核心代码为 poetry_generation_tf2.ipynb。

（1）进入 Anaconda Prompt 命令行。

（2）在本章根目录下输入命令 jupyter notebook。

（3）在浏览器中打开 poetry_generation_tf2.ipynb。

（4）分步运行代码，即可得到程序运行结果。

8.5.3 代码解析

（1）加载古诗数据集。

```
data_dir = './data/newtxt.txt'   #new_poetry
#text = open(data_dir, 'rb').read().decode(encoding='utf-8')text = open(data_dir, 'rb').read().decode(encoding='gb18030','ignore')
```

下面函数用来处理初始数据集 poetry.txt，当使用 newtxt.txt 这个文件时，可以不调用下面的程序代码。

```
import os
import re
pattern = '[a-zA-Z0-9'"#$%&\'()*+-./:;<=>@★…【】《》""''[\\]^_`{|}~]+'
def preprocess_poetry(outdir, datadir):
    with open(os.path.join(outdir, 'new_poetry.txt'), 'w') as out_f:
        with open(os.path.join(datadir, 'poetry.txt'), 'r') as f:
            for line in f:
                content = line.strip().rstrip('\n').split(':')[1]  #.rstrip('\n').
                content = content.replace(' ','')
                if '】' in content or '_' in content or '(' in content or '（' in content or '《' in content or '[' in content:
                    continue
                if len(content) < 20:
                    continue
                content=re.sub(pattern, '', content)
                out_f.write(content + '\n')
preprocess_poetry('./data/', './data/')
```

（2）实现数据预处理，需要先准备好汉字和 ID 之间的转换关系。在这个函数中，创建并返回两个字典：汉字到 ID 的转换字典 vocab_to_int；ID 到汉字的转换字典 int_to_vocab。

```
import numpy as np
from collections import Counter
import pickle
def create_lookup_tables():
    vocab = sorted(set(text))
    vocab_to_int = {u:i for i, u in enumerate(vocab)}
    int_to_vocab = np.array(vocab)
    int_text = np.array([vocab_to_int[word] for word in text if word != '\n'])
    pickle.dump((int_text, vocab_to_int, int_to_vocab), open('preprocess.p', 'wb'))
```

(3) 处理所有数据并保存,将每期结果按照从第一期开始的顺序保存到文件中。

```
create_lookup_tables()
import numpy as np
#读取保存的数据
int_text, vocab_to_int, int_to_vocab = pickle.load(open('preprocess.p', mode=
'rb'))
def get_batches(int_text, batch_size, seq_length):
    batchCnt = len(int_text) // (batch_size * seq_length)
    int_text_inputs = int_text[:batchCnt * (batch_size * seq_length)]
    int_text_targets = int_text[1:batchCnt * (batch_size * seq_length)+1]
    result_list = []
    x = np.array(int_text_inputs).reshape(1, batch_size, -1)
    y = np.array(int_text_targets).reshape(1, batch_size, -1)
    x_new = np.dsplit(x, batchCnt)
    y_new = np.dsplit(y, batchCnt)
    for ii in range(batchCnt):
        x_list = []
        x_list.append(x_new[ii][0])
        x_list.append(y_new[ii][0])
        result_list.append(x_list)
    return np.array(result_list)
```

(4) 训练 RNN 神经网络,超参数如下。

```
vocab_size = len(int_to_vocab)
#批次大小
batch_size = 32   #64
#RNN 的大小(隐藏节点的维度)
rnn_size = 1000
#嵌入层的维度
embed_dim = 256    #这里做了调整,跟彩票预测的也不同了
#序列的长度
seq_length = 15   #注意这里已经不是 1 了,在古诗预测里面这个数值可以大一些,如 100 也可以
save_dir = './save'
```

(5) 构建计算图,使用实现的神经网络构建计算图。

```
import tensorflow as tf
import datetime
from tensorflow import keras
from tensorflow.python.ops import summary_ops_v2
import time
physical_devices = tf.config.experimental.list_physical_devices('GPU')
assert len(physical_devices) > 0, 'Not enough GPU hardware devices available'
tf.config.experimental.set_memory_growth(physical_devices[0], True)
MODEL_DIR = "./poetry_models"
```

```python
train_batches = get_batches(int_text, batch_size, seq_length)
losses = {'train': [], 'test': []}
class poetry_network(object):
    def __init__(self, batch_size=32):
        self.batch_size = batch_size
        self.best_loss = 9999
        self.model = tf.keras.Sequential([
            tf.keras.layers.Embedding(vocab_size, embed_dim,
                                      batch_input_shape=[batch_size, None]),
            tf.keras.layers.LSTM(rnn_size,
                                 return_sequences=True,
                                 stateful=True,
                                 recurrent_initializer='glorot_uniform'),
            tf.keras.layers.Dense(vocab_size)
        ])
        self.model.summary()
        self.optimizer = tf.keras.optimizers.Adam()
        self.ComputeLoss = tf.keras.losses.SparseCategoricalCrossentropy(from_logits=True)
        if tf.io.gfile.exists(MODEL_DIR):
            #            print('Removing existing model dir: {}'.format(MODEL_DIR))
            #            tf.io.gfile.rmtree(MODEL_DIR)
            pass
        else:
            tf.io.gfile.makedirs(MODEL_DIR)
        train_dir = os.path.join(MODEL_DIR, 'summaries', 'train')
        self.train_summary_writer = summary_ops_v2.create_file_writer(train_dir, flush_millis=10000)
        checkpoint_dir = os.path.join(MODEL_DIR, 'checkpoints')
        self.checkpoint_prefix = os.path.join(checkpoint_dir, 'ckpt')
        self.checkpoint = tf.train.Checkpoint(model=self.model, optimizer=self.optimizer)
        #如果存在检查点,则在创建时恢复变量
        self.checkpoint.restore(tf.train.latest_checkpoint(checkpoint_dir))
    @tf.function
    def train_step(self, x, y):
        #记录计算损失所用的操作,使梯度可以计算有关变量的损失
        with tf.GradientTape() as tape:
            logits = self.model(x, training=True)
            loss = self.ComputeLoss(y, logits)
        grads = tape.gradient(loss, self.model.trainable_variables)
        self.optimizer.apply_gradients(zip(grads, self.model.trainable_variables))
        return loss, logits
    def training(self, epochs=1, log_freq=50):
        batchCnt = len(int_text) // (batch_size * seq_length)
```

```
            print("batchCnt : ", batchCnt)
        for i in range(epochs):
            train_start = time.time()
            with self.train_summary_writer.as_default():
                start = time.time()
                #Metrics are stateful. They accumulate values and return a cumulative
                # result when you call .result(). Clear accumulated values with .reset_
states()
                avg_loss = tf.keras.metrics.Mean('loss', dtype=tf.float32)
                #数据集可以像任何其他Python程序迭代一样被遍历
                for batch_i, (x, y) in enumerate(train_batches):
                    loss, logits = self.train_step(x, y)
                    avg_loss(loss)
                    losses['train'].append(loss)
                    if tf.equal(self.optimizer.iterations % log_freq, 0):
                        summary_ops_v2.scalar('loss', avg_loss.result(), step=
self.optimizer.iterations)
                        rate = log_freq / (time.time() - start)
                        print('Step #{}\tLoss: {:0.6f} ({} steps/sec)'.format(
                            self.optimizer.iterations.numpy(), loss, rate))
                        avg_loss.reset_states()
                        start = time.time()
#                        self.checkpoint.save(self.checkpoint_prefix)
            self.checkpoint.save(self.checkpoint_prefix)
            print("save model\n")
```

8.5.4 训练测试

对数据进行预处理后，开始训练神经网络，得到模型结构。

```
net = poetry_network()
net.training(20)
Model: "sequential"
_____
Layer (type)              Output Shape           Param #
=========================================================
embedding (Embedding)     (32, None, 256)        932352

lstm (LSTM)               (32, None, 1000)       5028000

dense (Dense)             (32, None, 3642)       3645642
=========================================================
Total params: 9,605,994
Trainable params: 9,605,994
Non-trainable params: 0
_____
```

```
batchCnt :    161
Step #2900   Loss: 1.443649 (945.9370954573953 steps/sec)
Step #2950   Loss: 1.249750 (40.23442093972123 steps/sec)
Step #3000   Loss: 1.260852 (40.06896636055711 steps/sec)
Step #3050   Loss: 1.000135 (40.00265520553809 steps/sec)
save model
Step #3100   Loss: 1.068481 (48.768288817569825 steps/sec)
Step #3150   Loss: 0.940284 (40.358050222020545 steps/sec)
Step #3200   Loss: 1.030366 (39.79661739986193 steps/sec)
save model
```

在训练过程中，训练损失代码如下所示，损失率随着轮次增加的变化如图 8-13 所示。

```
%matplotlib inline
%config InlineBackend.figure_format = 'retina'
# import seaborn as sns
import matplotlib.pyplot as plt
plt.plot(losses['train'], label='Training loss')
plt.legend()
_ = plt.ylim()
```

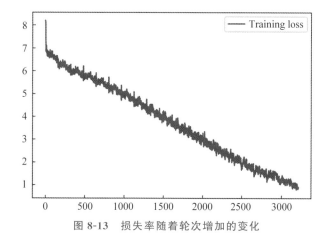

图 8-13　损失率随着轮次增加的变化

加载保存的模型准备预测。

```
restore_net=poetry_network(1)
restore_net.model.build(tf.TensorShape([1, None]))
Model: "sequential_1"
```

Layer (type)	Output Shape	Param #
embedding_1 (Embedding)	(1, None, 256)	932352
lstm_1 (LSTM)	(1, None, 1000)	5028000

```
dense_1 (Dense)              (1, None, 3642)            3645642
=================================================
Total params: 9,605,994
Trainable params: 9,605,994
Non-trainable params: 0
```

开始生成古诗。

参数含义如下。

(1) prime_word：开始的头一个字。

(2) top_n：从前 N 个候选汉字中随机选择。

(3) rule：默认是七言绝句。

(4) sentence_lines：生成几句古诗，默认是 4 句（逗号和句号结尾都算一句）。

(5) hidden_head：藏头诗的前几个字。

```python
def gen_poetry(prime_word='白', top_n=5, rule=7, sentence_lines=4, hidden_head=None):
    gen_length = sentence_lines * (rule + 1) - len(prime_word)
    gen_sentences = [prime_word] if hidden_head==None else [hidden_head[0]]
    temperature = 1.0
    dyn_input = [vocab_to_int[s] for s in prime_word]
    dyn_input = tf.expand_dims(dyn_input, 0)
    dyn_seq_length = len(dyn_input[0])
    restore_net.model.reset_states()
    index=len(prime_word) if hidden_head==None else 1
    for n in range(gen_length):
        index += 1
        predictions = restore_net.model(np.array(dyn_input))
        predictions = tf.squeeze(predictions, 0)
        if index!=0 and (index % (rule+1)) == 0:
            if ((index / (rule+1)) + 1) % 2 == 0:
                predicted_id=vocab_to_int[',']
            else:
                predicted_id=vocab_to_int['。']
        else:
            if hidden_head != None and (index-1)%(rule+1)==0 and (index-1)//(rule+1) < len(hidden_head):
                predicted_id=vocab_to_int[hidden_head[(index-1)//(rule+1)]]
            else:
                while True:
                    predictions = predictions / temperature
                    predicted_id = tf.random.categorical(predictions, num_samples=1)[-1, 0].numpy()
                    #p = np.squeeze(predictions[-1].numpy())
                    #p[np.argsort(p)[:-top_n]] = 0
                    #p = p / np.sum(p)
```

```
                    #c = np.random.choice(vocab_size, 1, p=p)[0]
                    #predicted_id=c
                    if(predicted_id != vocab_to_int[','] and predicted_id != vocab
_to_int['。']):
                        break
    #使用多项分布来预测模型返回的单词
    #        predictions = predictions / temperature
    #        predicted_id = tf.multinomial(predictions, num_samples=1)[-1,0].
numpy()
        dyn_input = tf.expand_dims([predicted_id], 0)
        gen_sentences.append(int_to_vocab[predicted_id])
    poetry_script = ' '.join(gen_sentences)
    poetry_script = poetry_script.replace('\n ', '\n')
    poetry_script = poetry_script.replace('( ', '(')
return poetry_script
```

通过运行上面的命令得到一些训练结果如图 8-14 所示。

给定开头

```
In  [14]: gen_poetry(prime_word='白日依山尽', top_n=10, rule=5, sentence_lines=4)
Out[14]: '白日依山尽 , 照 岸 终 年 矜 。 目 尽 飞 猿 笑 , 愁 长 断 独 声 。'
```

七言绝句

```
In  [168]: gen_poetry(prime_word='春', top_n=10, rule=7, sentence_lines=4)
Out[168]: '春 至 处 马 上 长 杨 , 谁 家 窈 窕 九 重 闺 。 秦 国 陇 台 汉 边 风 , 日 照 胡 沙 尘 花 中 。'
```

五言

```
In  [198]: gen_poetry(prime_word='月', top_n=10, rule=5, sentence_lines=4)
Out[198]: '月 来 长 河 闭 , 苔 疏 连 叶 满 。 归 来 入 汉 使 , 居 带 天 门 台 。'
```

藏头诗

```
In  [192]: gen_poetry(prime_word='夏', top_n=10, rule=7, sentence_lines=4, hidden_head='花好月圆')
Out[192]: '花襄茗蹇莽觭聂，好噸偷逮噸鲤蜘。月摁拖鸷标拖鸷，圆偷选偷袄鸷莽。'
```

图 8-14　一些训练结果

从训练结果上看，由于 RNN 对具有序列特性的数据非常有效，它能挖掘数据中的时序信息以及语义信息。因此，基于 RNN 的古诗生成模型可以较好地生成语义较为完整的目的诗词。

第 9 章

情感分类、翻译、对话

承接第 8 章的自然语言初步处理,本章将对自然语言做进一步探索。深度学习训练不仅可以对自然语言进行创作、加工,还可以对文本表达出的情感进行学习分类,计算文本之间的相关联程度。在充分的训练之后,相关模型还可以在不同语言之间进行翻译、聊天,从而实现人工智能更高级的应用。

9.1 使用双向循环神经网络对电影评论情感分类

本次实战中将实现一个堆叠的长短期记忆神经网络,用于 LSTM 文本数据集的情感分析训练并计算损失函数,最后验证结果[42]。

9.1.1 背景原理

IMDB(Internet Movie DataBase)是一个关于电影演员、电影、电视节目、电视明星和电影制作的在线数据库,其中包括影片的众多信息,如演员、片场、内容介绍、分级、评论等,截至 2021 年 6 月,IMDB 共收录了 800 多万部作品资料以及 1000 万名人物资料。

长短期记忆神经网络是一种特殊的循环神经网络,有效地解决了随训练时间增加以及网络层数的增多,原始 RNN 容易出现梯度爆炸或梯度消失的问题,主要应用于文本生成、机器翻译、语音识别、生成图像描述和视频标记等领域。

本章采用的是堆叠的长短期记忆神经网络,即将多层 LSTM 叠加起来形成多个隐藏层,图如图 9-1 所示。

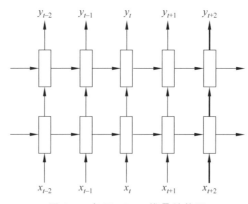

图 9-1 多层 LSTM 堆叠结构图

图 9-1 中每个 LSTM 单元内部结构图如图 9-2 所示,对于 l 层 t 时刻来说,h_{t-1}^{l} 为 l 层 $t-1$ 时刻(即上一个时刻)的输出,h_{t}^{l-1} 为 $l-1$ 层(即上一层)t 时刻的输出,这两个输出叠加作为 l 层 t 时刻的输入。

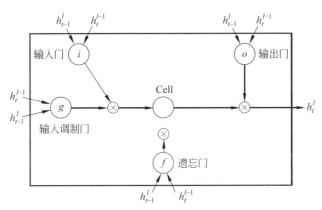

图 9-2　LSTM 单元内部结构图

9.1.2　安装操作

本节所用代码文件结构如表 9-1 所示。

表 9-1　代码文件结构

文 件 名 称	实 现 功 能
main.py	加载 IMDB 数据并训练

本次实战只有 main.py 这一个代码文件,在这个代码中对 IMDB 数据集进行训练。
安装步骤如下。
(1) 连接互联网(IMDB 数据集将从互联网上自动下载)。
(2) 进入 Anaconda Prompt 命令行,切换至本章根目录。
(3) 进入本节代码的目录。
(4) 在命令行中输入 python main.py。
(5) 运行代码,查看运行结果。

9.1.3　代码解析

下面讲解核心代码 main.py 的运行机制。

1. 导入函数库

```
import   os
import   tensorflow as tf
import   numpy as np
from     tensorflow import keras
tf.random.set_seed(22)
```

```python
np.random.seed(22)
os.environ['TF_CPP_MIN_LOG_LEVEL'] = '2'
assert tf.__version__.startswith('2.')
```

2. 修正随机种子并重新生成

```python
np.random.seed(7)
```

3. 加载数据时仅加载前 N 个词语，其他设置为 0

```python
top_words = 10000                    #编码10000个单词
```

4. 裁剪，用输入序列进行填补

```python
max_review_length = 80               #句子长度为80
(X_train, y_train), (X_test, y_test) = keras.datasets.imdb.load_data(num_words=top_words)
x_train = keras.preprocessing.sequence.pad_sequences(X_train, maxlen=max_review_length)
x_test = keras.preprocessing.sequence.pad_sequences(X_test, maxlen=max_review_length)
```

5. 递归神经网络定义

```python
class RNN(keras.Model):
    def __init__(self, units, num_classes, num_layers):
        super(RNN, self).__init__()
        self.rnn = keras.layers.LSTM(units, return_sequences=True)
        self.rnn2 = keras.layers.LSTM(units)
        self.embedding = keras.layers.Embedding(top_words, 100, input_length=max_review_length)
        self.fc = keras.layers.Dense(1)
    def call(self, inputs, training=None, mask=None):
        x = self.embedding(inputs)
        x = self.rnn(x)
        x = self.rnn2(x)
        x = self.fc(x)
        print(x.shape)
        return x
def main():
    units = 64
    num_classes = 2
    batch_size = 32
    epochs = 20
    model = RNN(units, num_classes, num_layers=2)
    model.compile(optimizer=keras.optimizers.Adam(0.001),
```

```
        loss=keras.losses.BinaryCrossentropy(from_logits=True),
        metrics=['accuracy'])
```

6．训练

```
model.fit(x_train, y_train, batch_size=batch_size, epochs=epochs,
          validation_data=(x_test, y_test), verbose=1)
```

7．在测试集上进行评估

```
    scores = model.evaluate(x_test, y_test, batch_size, verbose=1)
    print("Final test loss and accuracy :", scores)
if __name__ == '__main__':
    main()
```

9.1.4 训练测试

通过在命令行中运行 main.py，得到训练的轮次以及对应的损失率、准确率、验证数据的损失率、准确率。本次实战是一个深度学习训练阶段在情感分析方面的应用，图 9-3 直观解释了训练后的模型对不同电影的评论词汇热度分析的最终结果。

图 9-3　词汇热度分析最终结果

通过训练后得到的图像结果提取了英文词汇库的关键评论词的频率，其中词汇频率出现得越多，则在图中占据位置越接近图像布局的中间，占据的位置也越大。

9.2　应用词嵌入计算文本相关性

计算文本相关性的方法有很多，按照有无监督可分为无监督（Average Word Vector、Tfidf-Weighting Word Vectors、Bags of Words；LDA 等）和有监督（DSSM-LSTM、word2vec、doc2vec 等）。在本节中，将采用 word2vec 方法计算文本相关性，相比于其他方法来说，word2vec 是基于上下文内容进行预测，预测结果更好，而且维度较小计算速度更快，通用性也较强，可用于各种 NLP 任务中。

9.2.1 背景原理

自然语言是一套用来表达含义的复杂系统。在这套系统中,词是表义的基本单元。顾名思义,词向量是用来表示词的向量,也可被认为是词的特征向量或表征。把词映射为实数域向量的技术也叫词嵌入(Word Embedding)。近年来,词嵌入已逐渐成为自然语言处理的基础知识。

为何不采用 one-hot 向量呢?在前面的循环神经网络一节中使用 one-hot 向量表示词(字符为词)。回忆一下,假设词典中不同词的数量(词典大小)为 N,每个词可以和从 0 到 $N-1$ 的连续整数一一对应。这些与词对应的整数叫作词的索引。假设一个词的索引为 i,为了得到该词的 one-hot 向量表示,创建一个全 0 的长为 N 的向量,并将其第 i 位设成 1。这样一来,每个词就表示成了一个长度为 N 的向量,可以直接被神经网络使用。

虽然 one-hot 词向量构造起来很容易,但通常并不是一个好选择。一个主要的原因是,one-hot 词向量无法准确表达不同词之间的相似度,如常常使用的余弦相似度。由于任何两个不同词的 one-hot 向量的余弦相似度都为 0,多个不同词之间的相似度难以通过 one-hot 向量准确地体现出来。

word2vec 的出现正是为了解决上面这个问题。它将每个词表示成一个定长的向量,并使得这些向量能较好地表达不同词之间的相似和类比关系。word2vec 包含了两个模型,即跳字模型(Skip-Gram)和连续词袋模型(Continuous Bag of Words,CBOW)。跳字模型假设基于中心词来生成背景词。连续词袋模型假设基于背景词来生成中心词。

9.2.2 安装操作

本节所用代码文件结构如表 9-2 所示。

表 9-2 代码文件结构

文 件 名 称	实 现 功 能
word2vec-TensorFlow2.ipynb	加载 PTB 数据集并训练

本次实战只有 word2vec-Tensorflow2.ipynb 这一个代码文件,在这个代码中对 PTB 数据集进行训练。

安装步骤如下。

(1)进入 Anaconda Prompt 命令行,切换至本章根目录。
(2)在命令行中输入 jupyter notebook。
(3)在浏览器中打开 word2vec-Tensorflow2.ipynb。
(4)逐步运行代码,查看运行结果。

9.2.3 代码解析

1. 导入必备的库文件并处理数据集

PTB(Penn Tree Bank)是一个常用的小型语料库,它采样自《华尔街日报》的文章,包括训练集、验证集和测试集。下面将在 PTB 训练集上训练词嵌入模型。该数据集的每一行作为一个句子。句子中的每个词由空格隔开。对于数据集的前 3 个句子,打印每个句子的词数和前 5 个词。

```
import collections
import math
import random
import sys
import time
import os
import numpy as np
import tensorflow as tf
sys.path.append("..")
assert 'ptb.train.txt' in os.listdir("../../data/ptb")
with open('../../data/ptb/ptb.train.txt', 'r') as f:
    lines = f.readlines()
raw_dataset = [st.split() for st in lines]          #st 是 sentence 的缩写
'#sentences: %d' % len(raw_dataset)
for st in raw_dataset[:3]:
    print('#tokens:', len(st), st[:5])
```

2. 建立词语索引

为了计算简单,只保留在数据集中至少出现 5 次的词,然后将词映射到整数索引。

```
counter = collections.Counter([tk for st in raw_dataset for tk in st])
                                                    #tk 是 token 的缩写
counter = dict(filter(lambda x: x[1] >= 5, counter.items()))
idx_to_token = [tk for tk, _ in counter.items()]
token_to_idx = {tk: idx for idx, tk in enumerate(idx_to_token)}
dataset = [[token_to_idx[tk] for tk in st if tk in token_to_idx]
           for st in raw_dataset]
num_tokens = sum([len(st) for st in dataset])
'#tokens: %d' % num_tokens
```

3. 二次采样

文本数据中一般会出现一些高频词,如英文中的 the、a 和 in。通常来说,在一个背景窗口中,一个词(如 chip)和较低频词(如 microprocessor)同时出现比和较高频词(如 the)同时出现对训练词嵌入模型更有益。因此,训练词嵌入模型时可以对词进行二次采样。具体来说,数据集中每个被索引词 w_i 将有一定概率被丢弃。其中, $f(w_i)$ 是数据集中词 w_i 的个数与总词数之比,常数 t 是一个超参数。可见,只有当 $f(w_i) > t$ 时,才有可能在二次采样中丢弃词 w_i,并且越高频的词被丢弃的概率越大。

```
def discard(idx):
    return random.uniform(0, 1) < 1 - math.sqrt(
        1e-4 / counter[idx_to_token[idx]] * num_tokens)
```

```
subsampled_dataset = [[tk for tk in st if not discard(tk)] for st in dataset]
'#tokens: %d' % sum([len(st) for st in subsampled_dataset])
```

可以看到，二次采样后去掉了一半左右的词。下面比较一个词在二次采样前后出现在数据集中的次数。可见高频词 the 的采样率不足 1/20。

```
def compare_counts(token):
    return '#%s: before=%d, after=%d' % (token, sum(
        [st.count(token_to_idx[token]) for st in dataset]), sum(
        [st.count(token_to_idx[token]) for st in subsampled_dataset]))
compare_counts('the')
```

输出如下。

```
'#the: before=50770, after=2186'
```

但低频词 join 则完整地保留了下来。

```
compare_counts('join')
```

输出如下。

```
'#join: before=45, after=45'
```

4. 提取中心词和背景词

将与中心词距离不超过背景窗口大小的词作为它的背景词。下面定义函数提取出所有中心词和它们的背景词。它每次在整数 1 和 max_window_size（最大背景窗口）之间随机均匀采样一个整数作为背景窗口大小。

```
def get_centers_and_contexts(dataset, max_window_size):
    centers, contexts = [], []
    for st in dataset:
        if len(st) < 2:           #每个句子至少要有2个词才可能组成一对"中心词-背景词"
            continue
        centers += st
        for center_i in range(len(st)):
            window_size = random.randint(1, max_window_size)
            indices = list(range(max(0, center_i - window_size),
                                 min(len(st), center_i + 1 + window_size)))
            indices.remove(center_i)                    #将中心词排除在背景词之外
            contexts.append([st[idx] for idx in indices])
    return centers, contexts
```

下面创建一个人工数据集，其中含有词数分别为 7 和 3 的两个句子。设最大背景窗口为 2，打印所有中心词和它们的背景词。

```
tiny_dataset = [list(range(7)), list(range(7, 10))]
print('dataset', tiny_dataset)
for center, context in zip(*get_centers_and_contexts(tiny_dataset, 2)):
```

```
    print('center', center, 'has contexts', context)
```

输出如下。

```
dataset [[0, 1, 2, 3, 4, 5, 6], [7, 8, 9]]
center 0 has contexts [1, 2]
center 1 has contexts [0, 2, 3]
center 2 has contexts [0, 1, 3, 4]
center 3 has contexts [1, 2, 4, 5]
center 4 has contexts [2, 3, 5, 6]
center 5 has contexts [4, 6]
center 6 has contexts [4, 5]
center 7 has contexts [8, 9]
center 8 has contexts [7, 9]
center 9 has contexts [8]
```

实验中,设最大背景窗口大小为 5。下面提取数据集中所有的中心词及其背景词。

```
all_centers, all_contexts = get_centers_and_contexts(subsampled_dataset, 5)
```

5. 负采样

在此使用负采样来进行近似训练。对于一对中心词和背景词,随机采样 K 个噪声词(实验中设 $K=5$)。根据 word2vec 论文的建议,噪声词采样概率 $P(w)$ 设为 w 词频与总词频之比的 0.75 次方。根据每个词的权重(sampling_weights)随机生成 K 个词的索引作为噪声词。为了高效计算,可以将 K 设得稍大一点。

```
def get_negatives(all_contexts, sampling_weights, K):
    all_negatives, neg_candidates, i = [], [], 0
    population = list(range(len(sampling_weights)))
    for contexts in all_contexts:
        negatives = []
        while len(negatives) < len(contexts) * K:
            if i == len(neg_candidates):
                i, neg_candidates = 0, random.choices(
                    population, sampling_weights, k=int(1e5))
            neg, i = neg_candidates[i], i + 1
            #噪声词不能是背景词
            if neg not in set(contexts):
                negatives.append(neg)
        all_negatives.append(negatives)
    return all_negatives
sampling_weights = [counter[w] ** 0.75 for w in idx_to_token]
all_negatives = get_negatives(all_contexts, sampling_weights, 5)
```

6. 读取数据

从数据集中提取所有中心词 all_centers,以及每个中心词对应的背景词 all_contexts 和噪声词 all_negatives。将通过随机小批量来读取它们。

在一个小批量数据中，第 i 个样本包括一个中心词以及它所对应的 n_i 个背景词和 m_i 个噪声词。由于每个样本的背景窗口大小可能不一样，其中背景词与噪声词个数之和 n_i+m_i 也会不同。在构造小批量时，将每个样本的背景词和噪声词连接在一起，并添加填充项 0 直至连接后的长度相同，即长度均为 $\max(n_i+m_i)$（max_len 变量）。为了避免填充项对损失函数计算的影响，构造了掩码变量 masks，其每一个元素分别与连接后的背景词和噪声词 contexts_negatives 中的元素一一对应。当 contexts_negatives 变量中的某个元素为填充项时，相同位置的掩码变量 masks 中的元素取 0，否则取 1。为了区分正类和负类，还需要将 contexts_negatives 变量中的背景词和噪声词区分开来。依据掩码变量的构造思路，只需要创建与 contexts_negatives 变量形状相同的标签变量 labels，并将与背景词（正类）对应的元素设为 1，其余清 0。

下面实现这个小批量读取函数 batchify。它的小批量输入 data 是一个长度为批量大小的列表，其中每个元素分别包含中心词 center、背景词 context 和噪声词 negative。该函数返回的小批量数据符合需要的格式，例如，包含了掩码变量。

```
def batchify(data):
    max_len = max(len(c) + len(n) for _, c, n in data)
    centers, contexts_negatives, masks, labels = [], [], [], []
    for center, context, negative in data:
        center=center.numpy().tolist()
        context=context.numpy().tolist()
        negative=negative.numpy().tolist()
        cur_len = len(context) + len(negative)
        centers += [center]
        contexts_negatives += [context + negative + [0] * (max_len - cur_len)]
        masks += [[1] * cur_len + [0] * (max_len - cur_len)]
        labels += [[1] * len(context) + [0] * (max_len - len(context))]
    return tf.data.Dataset.from_tensor_slices((tf.reshape(tf.convert_to_tensor
(centers),shape=(-1, 1)), tf.convert_to_tensor(contexts_negatives),
            tf.convert_to_tensor(masks), tf.convert_to_tensor(labels)))
```

下面用刚刚定义的 batchify 函数指定 DataLoader 实例中小批量的读取方式，然后打印读取的第一个批量中各个变量的形状。

```
def generator():
    for cent, cont, neg in zip(all_centers,all_contexts,all_negatives):
        yield (cent, cont, neg)
batch_size = 512
dataset=tf.data.Dataset.from_generator(generator=generator,output_types=(tf.
int32,tf.int32, tf.int32))
dataset = dataset.apply(batchify).shuffle(len(all_centers)).batch(batch_size)
for batch in dataset:
    for name, data in zip(['centers', 'contexts_negatives', 'masks',
                    'labels'], batch):
```

```
        print(name, 'shape:', data.shape)
    break
```

输出如下。

```
centers shape: (512, 1)
contexts_negatives shape: (512, 60)
masks shape: (512, 60)
labels shape: (512, 60)
```

7. 嵌入层

获取词嵌入的层称为嵌入层，在 Keras 中可以通过创建 layers.Embedding 实例得到。嵌入层的权重是一个矩阵，其行数为词典大小（input_dim），列数为每个词向量的维度（output_dim）。设词典大小为 20，词向量的维度为 4。

```
embed = tf.keras.layers.Embedding(input_dim=20, output_dim=4)
embed.build(input_shape=(1,20))
embed.get_weights()
```

输出如下。

```
[array([[ 0.00186051, -0.02967212,  0.01367206, -0.00120816],
        [-0.02516781,  0.04460483,  0.0353159 , -0.00299061],
        [-0.01710566,  0.02493416, -0.04021205,  0.03587193],
        [ 0.02083813, -0.04663629,  0.03478238, -0.01295733],
        [ 0.01544365, -0.01052438,  0.0161058 , -0.02668259],
        [ 0.04031085, -0.0038617 ,  0.02243182,  0.01726558],
        [ 0.03033152, -0.00653177,  0.04465215,  0.04541362],
        [ 0.03119953,  0.04776872, -0.0443038 , -0.04148095],
        [-0.00150775,  0.02733872,  0.04239938,  0.01183169],
        [ 0.00960425,  0.00693228, -0.02657985, -0.04833677],
        [ 0.02114216,  0.04928992,  0.02301376, -0.00982954],
        [-0.0481813 , -0.00836793,  0.04207878, -0.04183043],
        [ 0.03138346, -0.02859619,  0.01237852, -0.03214811],
        [ 0.0256073 ,  0.00722118,  0.02427341, -0.04318798],
        [ 0.02297343,  0.0076545 , -0.04061831, -0.02541875],
        [ 0.00131279, -0.02258632,  0.04780482,  0.03646753],
        [-0.02359685, -0.0455141 , -0.02238152,  0.01958317],
        [-0.01812546,  0.00110344,  0.01814773, -0.01437292],
        [ 0.01395816, -0.04358598,  0.03148863,  0.03027416],
        [ 0.02516505,  0.0178661 , -0.02972771, -0.04738807]],
       dtype=float32)]
```

嵌入层的输入为词的索引。输入一个词的索引 i，嵌入层返回权重矩阵的第 i 行作为它的词向量。下面将形状为 (2, 3) 的索引输入嵌入层，由于词向量的维度为 4，得到形状为 (2, 3, 4) 的词向量。

```
x = tf.convert_to_tensor([[1, 2, 3], [4, 5, 6]], dtype=tf.float32)
embed(x)
```

输出如下。

```
<tf.Tensor: id=2251783, shape=(2, 3, 4), dtype=float32, numpy=
array([[[-0.02516781,  0.04460483,  0.0353159 , -0.00299061],
        [-0.01710566,  0.02493416, -0.04021205,  0.03587193],
        [ 0.02083813, -0.04663629,  0.03478238, -0.01295733]],
       [[ 0.01544365, -0.01052438,  0.0161058 , -0.02668259],
        [ 0.04031085, -0.0038617 ,  0.02243182,  0.01726558],
        [ 0.03033152, -0.00653177,  0.04465215,  0.04541362]]],
      dtype=float32)>
```

8. 小批量乘法

可以使用小批量乘法运算 batch_dot 对两个小批量中的矩阵一一做乘法。假设第一个小批量中包含 n 个形状为 $a \times b$ 的矩阵 $\boldsymbol{X}_1, \boldsymbol{X}_2 \cdots, \boldsymbol{X}_n$，第二个小批量中包含 n 个形状为 $b \times c$ 的矩阵 $\boldsymbol{Y}_1, \boldsymbol{Y}_2, \cdots, \boldsymbol{Y}_n$。这两个小批量的矩阵乘法输出为 n 个形状为 $a \times c$ 的矩阵 $\boldsymbol{X}_1\boldsymbol{Y}_1$，$\boldsymbol{X}_2\boldsymbol{Y}_2, \cdots, \boldsymbol{X}_n\boldsymbol{Y}_n$。因此，给定两个形状分别为 (n, a, b) 和 (n, b, c) 的 NDArray，小批量乘法输出的形状为 (n, a, c)。

```
X = tf.ones((2, 1, 4))
Y = tf.ones((2, 4, 6))
tf.matmul(X, Y).shape
```

输出如下。

```
TensorShape([2, 1, 6])
```

9. 跳字模型前向计算

在前向计算中，跳字模型的输入包含中心词索引 center 以及连接的背景词与噪声词索引 contexts_and_negatives。其中 center 变量的形状为（批量大小，1），而 contexts_and_negatives 变量的形状为（批量大小，'max_len'）。这两个变量先通过词嵌入层分别由词索引变换为词向量，再通过小批量乘法得到形状为（批量大小，1，'max_len'）的输出。输出中的每个元素是中心词向量与背景词向量或噪声词向量的内积。

```
def skip_gram(center, contexts_and_negatives, embed_v, embed_u):
    v = embed_v(center)
    u = embed_u(contexts_and_negatives)
    pred = tf.matmul(v, tf.transpose(u,perm=[0,2,1]))
    return pred
```

训练模型，二元交叉熵损失函数：根据负采样中损失函数的定义，可以直接使用 Keras 的二元交叉熵损失函数 SigmoidBinaryCrossEntropyLoss。

```
class SigmoidBinaryCrossEntropyLoss(tf.keras.losses.Loss):
    def __init__(self):           #none mean sum
        super(SigmoidBinaryCrossEntropyLoss, self).__init__()
    def __call__(self, inputs, targets, mask=None):
        #TensorFlow 中使用 tf.nn.weighted_cross_entropy_with_logits 设置 mask 并没
有起到作用
        #直接与mask按元素相乘会实现当mask为0时不计损失的效果
        inputs=tf.cast(inputs,dtype=tf.float32)
        targets=tf.cast(targets,dtype=tf.float32)
        mask=tf.cast(mask,dtype=tf.float32)
        res=tf.nn.sigmoid_cross_entropy_with_logits(inputs, targets) * mask
        return tf.reduce_mean(res,axis=1)
loss = SigmoidBinaryCrossEntropyLoss()
```

值得一提的是,可以通过掩码变量指定小批量中参与损失函数计算的部分预测值和标签:当掩码为 1 时,相应位置的预测值和标签将参与损失函数的计算;当掩码为 0 时,相应位置的预测值和标签则不参与损失函数的计算。之前提到,掩码变量可用于避免填充项对损失函数计算的影响。

```
pred = tf.convert_to_tensor([[1.5, 0.3, -1, 2], [1.1, -0.6, 2.2, 0.4]],dtype=tf.
float32)
#标签变量label中的1和0分别代表背景词和噪声词
label = tf.convert_to_tensor([[1, 0, 0, 0], [1, 1, 0, 0]],dtype=tf.float32)
mask = tf.convert_to_tensor([[1, 1, 1, 1], [1, 1, 1, 0]],dtype=tf.float32)
                                                                          #掩码变量
loss(label, pred, mask) * mask.shape[1] / tf.reduce_sum(mask,axis=1)
```

输出如下。

```
0.8740
1.2100
```

10. 初始化模型参数

分别构造中心词和背景词的嵌入层,并将超参数词向量维度 embed_size 设置成 100。

```
embed_size = 100
net = tf.keras.Sequential([
    tf.keras.layers.Embedding(input_dim=len(idx_to_token), output_dim=embed_
size),
    tf.keras.layers.Embedding(input_dim=len(idx_to_token), output_dim=embed_
size)
])
net.get_layer(index=0)
```

输出如下。

```
<tensorflow.python.keras.layers.embeddings.Embedding at 0xb449db3c8>
```

11. 定义训练函数

下面定义训练函数。由于填充项的存在，与之前的训练函数相比，损失函数的计算稍有不同。

```
def train(net, lr, num_epochs):
    optimizer = tf.keras.optimizers.Adam(learning_rate=lr)
    for epoch in range(num_epochs):
        start, l_sum, n = time.time(), 0.0, 0
        for batch in dataset:
            center, context_negative, mask, label = [d for d in batch]
            mask=tf.cast(mask,dtype=tf.float32)
            with tf.GradientTape(persistent=True) as tape:
                pred = skip_gram(center, context_negative, net.get_layer(index=0), net.get_layer(index=1))
                #使用掩码变量mask来避免填充项对损失函数计算的影响
                l = (loss(label, tf.reshape(pred,label.shape), mask) *
                    mask.shape[1] / tf.reduce_sum(mask,axis=1))
                l=tf.reduce_mean(l)                              #一个批次的平均损失
            
            grads = tape.gradient(l, net.variables)
            optimizer.apply_gradients(zip(grads, net.variables))
            l_sum += np.array(l).item()
            n += 1
        print('epoch %d, loss %.2f, time %.2fs'
              % (epoch + 1, l_sum / n, time.time() - start))
```

9.2.4 训练测试

执行下面的命令：train(net，0.01，10)。输出为每个轮次的损失函数和所需要花费的时间。

```
epoch 1, loss 0.44, time 60.71s
epoch 2, loss 0.38, time 72.93s
epoch 3, loss 0.34, time 65.19s
epoch 4, loss 0.32, time 60.64s
epoch 5, loss 0.31, time 59.66s
epoch 6, loss 0.30, time 60.31s
epoch 7, loss 0.30, time 60.42s
epoch 8, loss 0.29, time 59.51s
epoch 9, loss 0.29, time 59.61s
epoch 10, loss 0.28, time 63.11s
def get_similar_tokens(query_token, k, embed):
    W = embed.get_weights()
    W = tf.convert_to_tensor(W[0])
```

```
        x = W[token_to_idx[query_token]]
        x = tf.reshape(x,shape=[-1,1])
        #添加的 1e-9 是为了数值稳定性
        cos = tf.reshape(tf.matmul(W, x),shape=[-1])/ tf.sqrt(tf.reduce_sum(W * W,
    axis=1) * tf.reduce_sum(x * x) + 1e-9)
        _, topk = tf.math.top_k(cos, k=k+1)
        topk=topk.numpy().tolist()
        for i in topk[1:]:              #除去输入词
            print('cosine sim=%.3f: %s' % (cos[i], (idx_to_token[i])))
    get_similar_tokens('chip', 3, net.get_layer(index=0))
```

应用词嵌入模型的输出为相似的关联词汇。

```
cosine sim=0.539: nec
cosine sim=0.494: a.g.
cosine sim=0.493: microprocessor
cosine sim=0.594: microprocessor
cosine sim=0.567: intel
cosine sim=0.524: workstations
```

可以使用其通过负采样训练跳字模型。二次采样试图尽可能减轻高频词对训练词嵌入模型的影响。可以将长度不同的样本填充至长度相同的小批量，并通过掩码变量区分非填充和填充，然后只令非填充参与损失函数的计算。

负采样通过考虑同时含有正类样本和负类样本的相互独立事件来构造损失函数。其训练中每一步的梯度计算开销与采样的噪声词的个数线性相关。层序 softmax 使用了二叉树，并根据根节点到叶节点的路径来构造损失函数。其训练中每一步的梯度计算开销与词典大小的对数相关。

9.3 中英翻译机器人

本次实战视图通过深度学习训练一个中英翻译机器人。该机器人主要基于来自 Transformer 模型的双向编码器表示（BERT）。BERT 是一种基于 Transformer 模型的机器学习技术，用于自然语言处理（NLP）。

BERT 对于 NLP 来说具有非常重要的意义，大部分 NLP 任务都存在着数据少、训练模型浅的问题，而 BERT 利用 Transformer 模型的编码器通过在语言模型上做预训练[43]。

9.3.1 背景原理

BERT 本质上是通过在海量语料的基础上运行自监督学习方法为单词学习一个好的特征表示，所谓自监督学习，是指在没有人工标注的数据上运行的监督学习。在以后特定的 NLP 任务中，可以直接使用 BERT 的特征表示作为该任务的词嵌入特征。所以，BERT 提供的是一个供其他任务迁移学习的模型，该模型可以根据任务微调或者固定之后作为特征提取器。

BERT 的输入向量编码（长度是 512）是 3 个嵌入特征的单位和，如图 9-4 所示。

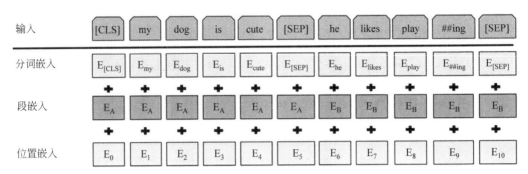

图 9-4　BERT 的输入向量编码

预训练时，BERT 由有两个自监督任务 MLM 和 NSP 组成，前者是指在训练时随机从输入语料上遮挡一些单词，然后通过上下文预测该单词；后者是判断句子 B 是否是句子 A 的下文，若是则输出 IsNext，若不是则输出 NotNext。当训练完后便可将其应用到 NLP 的各个任务中。

9.3.2　安装操作

本节所用代码文件结构如表 9-3 所示。

表 9-3　代码文件结构

文 件 名 称	实 现 功 能	文 件 名 称	实 现 功 能
transformer_train.py	文本翻译训练文件	bert_train.py	文本翻译测试文件

安装步骤如下。

（1）安装所需的软件包。

（2）在命令行输入 python transformer_training.py，运行 transformer 模型等进行训练。

（3）在命令行输入 python bert_train.py，运行 BERT 预先训练好的编码器模型。

9.3.3　代码解析

首先介绍第一个程序 transformer_train.py。

（1）导入函数库。

```
import   tensorflow as tf
import   time
import   numpy as np
import   matplotlib.pyplot as plt
import   os
```

下面的 tf.config.experimental.set_memory_growth 可以很好地将 GPU 的内存进行动态分配，避免了内存使用后不释放的问题，较好解决了内存溢出的问题。

```
os.environ['TF_CPP_MIN_LOG_LEVEL'] = '2'
```

```
gpus = tf.config.experimental.list_physical_devices('GPU')
if gpus:
  try:
    for gpu in gpus:
      tf.config.experimental.set_memory_growth(gpu, True)
    logical_gpus = tf.config.experimental.list_logical_devices('GPU')
    print(len(gpus), "Physical GPUs,", len(logical_gpus), "Logical GPUs")
  except RuntimeError as e:
print(e)
```

下面是导入 transformer 等必需的一些函数库。

```
from    tokenizer import get_tokenizer
from    transformer import Transformer
from    utils import CustomSchedule, create_masks
from    test import Translator
BUFFER_SIZE = 5000
BATCH_SIZE = 16
MAX_SEQ_LENGTH = 32
train_dataset, val_dataset, tokenizer_en, tokenizer_zh = \
    get_tokenizer(MAX_SEQ_LENGTH, BATCH_SIZE)
```

（2）下面的程序是中文向英文转换的 transformer 程序。

```
input_vocab_size = 21128
target_vocab_size = tokenizer_en.vocab_size + 2
dropout_rate = 0.1
num_layers=4
d_model=512
dff=2048
num_heads=8
transformer = Transformer(num_layers, d_model, num_heads, dff, input_vocab_size,
target_vocab_size, dropout_rate)
inp = tf.random.uniform((BATCH_SIZE, MAX_SEQ_LENGTH))
tar_inp = tf.random.uniform((BATCH_SIZE, MAX_SEQ_LENGTH))
fn_out, _ = transformer(inp, tar_inp,
                    True,
                    enc_padding_mask=None,
                    look_ahead_mask=None,
                    dec_padding_mask=None)
print(tar_inp.shape)            # (批大小，目标序列的长度)
print(fn_out.shape)             # (批大小，目标序列的长度，目标字典的大小)
transformer.summary()
learning_rate = CustomSchedule(d_model)
optimizer = tf.keras.optimizers.Adam(learning_rate, beta_1=0.9, beta_2=0.98,
epsilon=1e-9)
loss_object = tf.keras.losses.SparseCategoricalCrossentropy(from_logits=True,
```

```python
                reduction='none')
def loss_function(real, pred):
    mask = tf.math.logical_not(tf.math.equal(real, 0))
    loss_ = loss_object(real, pred)
    mask = tf.cast(mask, dtype=loss_.dtype)
    loss_ *= mask
    return tf.reduce_mean(loss_)
train_loss = tf.keras.metrics.Mean(name='train_loss')
train_accuracy = tf.keras.metrics.SparseCategoricalAccuracy(name='train_accuracy')
checkpoint_path = "./zh-en/transformer"
ckpt = tf.train.Checkpoint(transformer=transformer, optimizer=optimizer)
ckpt_manager = tf.train.CheckpointManager(ckpt, checkpoint_path, max_to_keep=5)
```

（3）训练并固化下来必要的模型，其中中文作为输入语言，英文作为输出语言。

```python
if ckpt_manager.latest_checkpoint:
    ckpt.restore(ckpt_manager.latest_checkpoint)
    print('Latest checkpoint restored!!')
@tf.function
def train_step(inp, tar):
    tar_inp = tar[:, :-1]
    tar_real = tar[:, 1:]
    enc_padding_mask, combined_mask, dec_padding_mask = create_masks(inp, tar_inp)
    with tf.GradientTape() as tape:
        predictions, _ = transformer(inp, tar_inp,
                                     True,
                                     enc_padding_mask,
                                     combined_mask,
                                     dec_padding_mask)
        loss = loss_function(tar_real, predictions)
    gradients = tape.gradient(loss, transformer.trainable_variables)
    optimizer.apply_gradients(zip(gradients, transformer.trainable_variables))
    train_loss(loss)
    train_accuracy(tar_real, predictions)
translator = Translator(tokenizer_zh, tokenizer_en, transformer, MAX_SEQ_LENGTH)
for epoch in range(20):
    (cn_code, en_code) = next(iter(val_dataset))
    cn_code, en_code = cn_code[epoch].numpy(), en_code[epoch].numpy()
    #print(cn_code)
    #print(en_code)
    en = tokenizer_en.decode([i for i in en_code if i < tokenizer_en.vocab_size])
    cn_code = [int(i)
               for i in cn_code if (i!=101 and i!=102 and i!=1 and i!=0)]
```

```
        #print(cn_code)
        cn = tokenizer_zh.convert_ids_to_tokens(cn_code)
        cn = "".join(cn)
        translator.do(cn)
        print('Real:', en)
        print('\n')
    start = time.time()
    train_loss.reset_states()
    train_accuracy.reset_states()
    #inp -> chinese, tar -> english
    for (batch, (inp, tar)) in enumerate(train_dataset):
        train_step(inp, tar)
        if batch % 50 == 0:
            print('Epoch {} Batch {} Loss {:.4f} Accuracy {:.4f}'.format(
                epoch + 1, batch, train_loss.result(), train_accuracy.result()))
    if (epoch + 1) % 3 == 0:
        ckpt_save_path = ckpt_manager.save()
        print('Saving checkpoint for epoch {} at {}'.format(epoch + 1,
                                                            ckpt_save_path))
    print('Epoch {} Loss {:.4f} Accuracy {:.4f}'.format(epoch + 1,
                                                       train_loss.result(),
                                                       train_accuracy.result()))
    print('Time taken for 1 epoch: {} secs\n'.format(time.time() - start))
```

bert_train.py 是测试模型的核心程序,具体解析如下。

(1) 导入相关 TensorFlow 的库函数,目前来说本程序的 GPU 的内存分配必须是保持一致的,并且在初始化 GPU 之前,要对内存增长进行设置,即动态内存分配。

```
import  tensorflow as tf
import  time
import  numpy as np
import  matplotlib.pyplot as plt
import  os
os.environ['TF_CPP_MIN_LOG_LEVEL'] = '2'
gpus = tf.config.experimental.list_physical_devices('GPU')
if gpus:
  try:
    for gpu in gpus:
      tf.config.experimental.set_memory_growth(gpu, True)
    logical_gpus = tf.config.experimental.list_logical_devices('GPU')
    print(len(gpus), "Physical GPUs,", len(logical_gpus), "Logical GPUs")
  except RuntimeError as e:
print(e)
```

(2) 下面的程序是借助 chinese_L-12_H-768_A-12 这个已经训练好的模型,完成中文向英文转换。

```python
from      tokenizer import get_tokenizer
from      bertmodel import Transformer, Config
from      utils import CustomSchedule, create_masks
from      test import Translator
BUFFER_SIZE = 50000
BATCH_SIZE = 64
MAX_SEQ_LENGTH = 128
train_dataset, val_dataset, tokenizer_en, tokenizer_zh = \
    get_tokenizer(MAX_SEQ_LENGTH, BATCH_SIZE)
config = Config(num_layers=6, d_model=256, dff=1024, num_heads=8)
target_vocab_size = tokenizer_en.vocab_size + 2
dropout_rate = 0.1
MODEL_DIR = "chinese_L-12_H-768_A-12"
bert_config_file = os.path.join(MODEL_DIR, "bert_config.json")
bert_ckpt_file = os.path.join(MODEL_DIR, "bert_model.ckpt")
transformer = Transformer(config=config,
                          target_vocab_size=target_vocab_size,
                          bert_config_file=bert_config_file)
inp = tf.random.uniform((BATCH_SIZE, MAX_SEQ_LENGTH))
tar_inp = tf.random.uniform((BATCH_SIZE, MAX_SEQ_LENGTH))
fn_out, _ = transformer(inp, tar_inp,
                        True,
                        enc_padding_mask=None,
                        look_ahead_mask=None,
                        dec_padding_mask=None)
print(tar_inp.shape)            # (批大小, 目标序列的长度)
print(fn_out.shape)             # (批大小, 目标序列的长度, 目标字典的大小)
```

初始化权重文件,并设计定义损失函数。

```python
transformer.restore_encoder(bert_ckpt_file)
transformer.summary()
learning_rate = CustomSchedule(config.d_model)
optimizer = tf.keras.optimizers.Adam(learning_rate, beta_1=0.9, beta_2=0.98,
                                    epsilon=1e-9)
loss_object = tf.keras.losses.SparseCategoricalCrossentropy(
    from_logits=True, reduction='none')
def loss_function(real, pred):
    mask = tf.math.logical_not(tf.math.equal(real, 0))
    loss_ = loss_object(real, pred)
    mask = tf.cast(mask, dtype=loss_.dtype)
    loss_ *= mask
    return tf.reduce_mean(loss_)
train_loss = tf.keras.metrics.Mean(name='train_loss')
train_accuracy = tf.keras.metrics.SparseCategoricalAccuracy(
    name='train_accuracy')
```

```python
checkpoint_path = "./zh-en/bert"
ckpt = tf.train.Checkpoint(transformer=transformer, optimizer=optimizer)
ckpt_manager = tf.train.CheckpointManager(ckpt, checkpoint_path, max_to_keep=5)
#如果模型存在,就直接利用现有模型进行训练
if ckpt_manager.latest_checkpoint:
    ckpt.restore(ckpt_manager.latest_checkpoint)
    print('Latest checkpoint restored!!')
@tf.function
def train_step(inp, tar):
    tar_inp = tar[:, :-1]
    tar_real = tar[:, 1:]
    enc_padding_mask, combined_mask, dec_padding_mask = create_masks(inp, tar_inp)
    with tf.GradientTape() as tape:
        predictions, _ = transformer(inp, tar_inp,
                                     True,
                                     enc_padding_mask,
                                     combined_mask,
                                     dec_padding_mask)
        loss = loss_function(tar_real, predictions)
    gradients = tape.gradient(loss, transformer.trainable_variables)
    optimizer.apply_gradients(zip(gradients, transformer.trainable_variables))
    train_loss(loss)
    train_accuracy(tar_real, predictions)
#下面是输入的中文,然后得到的英文翻译作为输出
translator = Translator(tokenizer_zh, tokenizer_en, transformer, MAX_SEQ_LENGTH)
for epoch in range(4):
    res = translator.do('周四和周五发生在5个省份的炸弹爆炸事件导致4人死亡,几十人受伤,遭袭地点包括旅游胜地普吉和华欣。')
    start = time.time()
    train_loss.reset_states()
    train_accuracy.reset_states()
    #inp -> chinese, tar -> english
    for (batch, (inp, tar)) in enumerate(train_dataset):
        train_step(inp, tar)
        if batch % 500 == 0:
            print('Epoch {} Batch {} Loss {:.4f} Accuracy {:.4f}'.format(
                epoch + 1, batch, train_loss.result(), train_accuracy.result()))
    if (epoch + 1) % 1 == 0:
        ckpt_save_path = ckpt_manager.save()
        print('Saving checkpoint for epoch {} at {}'.format(epoch + 1, ckpt_save_path))
    print('Epoch {} Loss {:.4f} Accuracy {:.4f}'.format(epoch + 1, train_loss.result(), train_accuracy.result()))
    print('Time taken for 1 epoch: {} secs\n'.format(time.time() - start))
```

9.3.4 训练测试

下面的例子就是输入中文后,通过预训练模型 BERT 后翻译出的结果。
中文输入如下。

Chinese src:周四和周五发生在 5 个省份的炸弹爆炸事件导致 4 人死亡,几十人受伤,遭袭地点包括旅游胜地普吉和华欣。

通过模型输出的结果如下。

Translated: Four weeks and five-week bombings have killed four people and injured dozens of others, including tourism and the likelihood of land grabs.

真实翻译结果如下。

Real: The bombings in five provinces on Thursday and Friday, including in Phuket and Hua Hin, areas popular with tourists, killed four people and injured dozens.

9.4 基于 Seq2Seq 中文聊天机器人实战

自然语言处理是人工智能应用较为成熟的领域,本节将通过中文聊天机器人这个项目介绍语言模型、循环神经网络和 Seq2Seq 模型。

9.4.1 背景原理

语言模型的发源历史久远,分为统计语言模型、n-gram 语言模型和神经网络语言模型 3 个阶段。统计语言模型是统计每个词出现的频次来形成词频字典,然后根据输入计算下一个输出词的概率,最后形成输出语句的。关于 n-gram 语言模型,利用统计语言模型计算输出语句概率的数量多到无法计算,这是由依据链式法则进行展开后全量连乘所引起的,那么解决这个问题的方法只有一个,就是缩短连乘的长度,其理论依据是马尔可夫假设。关于神经网络,是在 n-gram 语言模型的基础上改进的,采用 one-hot 编码表示每个词的分布情况,将输入语句进行编码转换后输入神经网络,经过 tanh 非线性变换和 Softmax 归一化后得到一个总和为 1 的向量,在向量中最大元素的下标作为输出词的字典编码,通过字典编码查询字典得到最终的输出词。以上过程一次可以得到一个输出词,如果要输出一句话以上就要循环以上过程,这就是接下来要讲的循环神经网络。

Seq2Seq(Sequence to Sequence)是基于 Encoder-Decoder 框架的 RNN 变种。Seq2Seq 引入了 Encoder-Decoder 框架,提高了神经网络对长文本信息的提取能力,取得了比单纯使用 LSTM 更好的效果,广泛应用于语言翻译和语言生成等 NLP 任务中。Seq2Seq 中有两个概念需要掌握:一个是 Encoder-Decoder 框架,另一个是 Attention 机制。

Encoder-Decoder 是处理输入、输出长短不一的多对多文本预测问题的框架,它提供了有效的文本特征提取、输出预测机制。Encoder-Decoder 框架包含两部分内容,分别是 Encoder 和 Decoder。

Encoder 的作用是对输入的文本信息进行有效的编码后将其作为 Decoder 的输入数

据，目的是对输入的文本信息进行特征提取，尽量准确高效地表征该文本的特征信息。

Decoder 的作用是从上下文的文本信息中获取尽可能多的特征，然后输出预测文本。根据文本信息的获取方式不同，Decoder 一般分为 4 种，分别是直译式解码、循环式解码、增强循环式解码和注意力机制解码。直译式解码是按照编码器的方式进行逆操作得到预测文本。循环式解码是将编码器输出的编码向量作为第一时刻的输入，然后将得到的输出作为下一个时刻的输入，以此进行循环解码。增强循环式解码是在循环式解码的基础上，每个时刻增加一个编码器输出的编码向量并以此作为输入。注意力机制解码是在增强循环式解码的基础上增加注意力机制，这样可以有效地训练解码器在繁多的输入中重点关注某些有效特征信息，以增加解码器的特征获取能力，进而得到更好的解码效果。

Attention 机制有效地解决了输入长序列信息时真实含义获取难的问题，在进行长序列处理的任务中，影响当前时刻状态的信息可能隐藏在前面的时刻里，根据马尔可夫假设这些信息有可能就会被忽略掉。举例说明，在"我快饿死了，我今天做了一天的苦力，我要大吃一顿"这句话中，可以明白"我要大吃一顿"是因为"我快饿死了"，但是根据马尔可夫假设，"我今天做了一天的苦力"和"我要大吃一顿"在时序上离得更近，相比于"我快饿死了"，"我今天做了一天的苦力"对"我要大吃一顿"的影响更强，但是在真实的自然语言中不是这样的。

从这个例子可以看出，神经网络模型没有很好地准确获取倒装时序的语言信息，要解决这个问题就需要经过训练自动建立起"我要大吃一顿"和"我快饿死了"的关联关系，这就是注意力机制。

本次实战为应用网络平台搭建一个聊天机器人。

9.4.2 安装操作

本节所用代码文件结构如表 9-4 所示。

表 9-4 代码文件结构

文 件 名 称	实 现 功 能	文 件 名 称	实 现 功 能
app.py	启动聊天机器人	execute.py	训练聊天机器人
data_utils.py	处理数据	getConfig.py	获取配置信息
seq2seqModel.py	定义 Seq2Seq 模型		

安装步骤如下。

（1）执行 python execute.py 进行模型的训练，为了简化训练配置，所有的训练设置成了死循环。当需要终止训练时，可以使用 Ctrl＋Z 键终止训练进程。当需要继续训练时，执行 python3 execute.py 会自动在之前训练结果的基础上继续训练。

（2）训练一段时间之后，可以执行 python app.py，在浏览器中访问提示的 URL 来体验训练效果。

9.4.3 代码解析

本次实战核心代码为 execute.py 和 app.py。

下面解析 execute.py。

```
#-*-coding:utf-8-*-
```

(1)加载 GPU 动态内存分配的函数库。

```
import os
import sys
import time
import tensorflow as tf
import seq2seqModel
import getConfig
import io
physical_devices = tf.config.experimental.list_physical_devices('GPU')
assert len(physical_devices) > 0, "Not enough GPU hardware devices available"
tf.config.experimental.set_memory_growth(physical_devices[0], True)
gConfig = {}
gConfig=getConfig.get_config(config_file='seq2seq.ini')
vocab_inp_size = gConfig['enc_vocab_size']
vocab_tar_size = gConfig['dec_vocab_size']
embedding_dim=gConfig['embedding_dim']
units=gConfig['layer_size']
BATCH_SIZE=gConfig['batch_size']
max_length_inp,max_length_tar=20,20
```

(2)定义数据预处理,读取数据,设置配置文件,确定训练函数。

```
def preprocess_sentence(w):
    w ='start '+ w + ' end'
    #print(w)
    return w
def create_dataset(path, num_examples):
    lines = io.open(path, encoding='UTF-8').read().strip().split('\n')
    word_pairs = [[preprocess_sentence(w) for w in l.split('\t')] for l in lines[:num_examples]]
    return zip(*word_pairs)
def max_length(tensor):
    return max(len(t) for t in tensor)
def read_data(path,num_examples):
    input_lang,target_lang=create_dataset(path,num_examples)
    input_tensor,input_token=tokenize(input_lang)
    target_tensor,target_token=tokenize(target_lang)
    return input_tensor,input_token,target_tensor,target_token
def tokenize(lang):
    lang_tokenizer = tf.keras.preprocessing.text.Tokenizer(num_words=gConfig['enc_vocab_size'], oov_token=3)
    lang_tokenizer.fit_on_texts(lang)
    tensor = lang_tokenizer.texts_to_sequences(lang)
    tensor = tf.keras.preprocessing.sequence.pad_sequences(tensor, maxlen=max_length_inp,padding='post')
```

```
    return tensor, lang_tokenizer
input_tensor,input_token,target_tensor,target_token= read_data(gConfig['seq_
data'], gConfig['max_train_data_size'])
def train():
    print("Preparing data in %s" % gConfig['train_data'])
    steps_per_epoch = len(input_tensor) // gConfig['batch_size']
    print(steps_per_epoch)
    enc_hidden = seq2seqModel.encoder.initialize_hidden_state()
    checkpoint_dir = gConfig['model_data']
ckpt=tf.io.gfile.listdir(checkpoint_dir)
```

(3)训练并保存检查点。

```
    if ckpt:
        print("reload pretrained model")

        seq2seqModel.checkpoint.restore(tf.train.latest_checkpoint(checkpoint_
dir))
    BUFFER_SIZE = len(input_tensor)
    dataset = tf.data.Dataset.from_tensor_slices((input_tensor, target_tensor)).
shuffle(BUFFER_SIZE)
    dataset = dataset.batch(BATCH_SIZE, drop_remainder=True)
    checkpoint_dir = gConfig['model_data']
    checkpoint_prefix = os.path.join(checkpoint_dir, "ckpt")
    start_time = time.time()
    while True:
        start_time_epoch = time.time()
        total_loss = 0
        for (batch, (inp, targ)) in enumerate(dataset.take(steps_per_epoch)):
            batch_loss = seq2seqModel.train_step(inp, targ, target_token, enc_
hidden)
            total_loss += batch_loss
            print(batch_loss.numpy())
        step_time_epoch = (time.time() - start_time_epoch) / steps_per_epoch
        step_loss = total_loss / steps_per_epoch
        current_steps = +steps_per_epoch
        step_time_total = (time.time() - start_time) / current_steps
        print('训练总步数：{} 每步耗时：{}   最新每步耗时：{} 最新每步 loss {:.4f}'.
format(current_steps, step_time_total, step_time_epoch,

step_loss.numpy()))
        seq2seqModel.checkpoint.save(file_prefix=checkpoint_prefix)
        sys.stdout.flush()
def predict(sentence):
    checkpoint_dir = gConfig['model_data']
     seq2seqModel.checkpoint.restore(tf.train.latest_checkpoint(checkpoint_
```

```
        dir))
        sentence = preprocess_sentence(sentence)
        inputs = [input_token.word_index.get(i,3) for i in sentence.split(' ')]
        inputs = tf.keras.preprocessing.sequence.pad_sequences([inputs],maxlen=max_length_inp,padding='post')
        inputs = tf.convert_to_tensor(inputs)
        result = ''
        hidden = [tf.zeros((1, units))]
        enc_out, enc_hidden = seq2seqModel.encoder(inputs, hidden)
        dec_hidden = enc_hidden
        dec_input = tf.expand_dims([target_token.word_index['start']], 0)
        for t in range(max_length_tar):
            predictions, dec_hidden, attention_weights = seq2seqModel.decoder(dec_input, dec_hidden, enc_out)
            predicted_id = tf.argmax(predictions[0]).numpy()
            if target_token.index_word[predicted_id] == 'end':
                break
            result += target_token.index_word[predicted_id] + ' '
            dec_input = tf.expand_dims([predicted_id], 0)
        return result
if __name__ == '__main__':
    if len(sys.argv) - 1:
        gConfig = getConfig.get_config(sys.argv[1])
    else:
        gConfig = getConfig.get_config()
    print('\n>> Mode : %s\n' % (gConfig['mode']))
    if gConfig['mode'] == 'train':
        train()
    elif gConfig['mode'] == 'serve':
        print('Serve Usage : >> python3 app.py')
```

下面为 app.py。

载入网络操作函数库和训练函数库并定义心跳检测函数用于定时刷新对话状态。

```
#coding=utf-8
from flask import Flask, render_template, request,jsonify
import execute
import time
import threading
import jieba
"""
定义心跳检测函数
"""
def heartbeat():
    print (time.strftime('%Y-%m-%d %H:%M:%S - heartbeat', time.localtime(time.time())))
```

```
    timer = threading.Timer(60, heartbeat)
    timer.start()
timer = threading.Timer(60, heartbeat)
timer.start()
"""
```

ElementTree 在 Python 标准库中有两种实现：一种是纯 Python 实现，例如 xml.etree.ElementTree；另一种是 C 语言实现，例如 xml.etree.cElementTree。

尽量使用第二种，因为它速度更快，而且消耗的内存更少。

```
"""
app = Flask(__name__,static_url_path="/static")
@app.route('/message', methods=['POST'])

#定义应答函数,用于获取输入信息并返回相应的答案
def reply():
#从请求中获取参数信息
    req_msg = request.form['msg']
#将语句使用结巴分词进行分词
    req_msg=" ".join(jieba.cut(req_msg))
    #调用 decode_line 对生成回答信息
    res_msg = execute.predict(req_msg)
    #将_UNK 值的词用微笑符号代替
    res_msg = res_msg.replace('_UNK', '^_^')
    res_msg=res_msg.strip()
    #如果接收到的内容为空,则给出相应的回复
    if res_msg == ' ':
      res_msg = '请与我聊聊天吧'
    return jsonify( { 'text': res_msg } )
"""
```

jsonify 是用于处理序列化 json 数据的函数，就是将数据组装成 json 格式返回。

http://flask.pocoo.org/docs/0.12/api/#module-flask.json

```
"""
@app.route("/")
def index():
    return render_template("index.html")
'''
'''
#启动 App
if (__name__ == "__main__"):
#    app.run(host = '0.0.0.0', port = 8808)
    app.run(debug=True)
```

9.4.4　训练测试

通过运行上面的 execute.py 训练得到检查点，再运行 app.py，最后加载模型文件打开

浏览器进入如图 9-5 所示聊天界面。

图 9-5　聊天界面

在对话框里输入聊天内容，单击 SEND 按钮后机器人就可以自动与你开始聊天了。

第 10 章

GAN 及其变体的创作

本章介绍生成对抗网络在 TensorFlow 实战中的应用。生成对抗网络的核心原理就是通过生成不同类型的数据进行预测，运用约束条件不断判别与实际的数据是否相符[44]。可以应用于相似动物的转换、漫画风格的互换、人脸生成和其他符合生成对抗原理的深度学习任务。

10.1 GAN 的原理与实战

"魔高一尺，道高一丈。"这句话形容生成对抗网络再合适不过了。本节实战主要通过对抗生成网络来训练生成漫画中的人脸[51]。

10.1.1 背景原理

GAN 会创建两个不同的对立网络，目的是让一个网络生成的样本与训练集不同，并且要让另外一个网络难辨真假。在"图灵测试"中，计算机试图与人对话，并让人误以为它也是一个正常人类。与此类似，在 GAN 中，生成器（Generator）负责生产作品，而鉴别器（Discriminator）负责甄别辨认。

假设 D 和 G 同步学习成长，G 向卡通名家 K 学习绘制人物头像，而 D 则学习如何鉴别作品的真伪。刚开始时 G 只学会了简单的笔墨用法，作品如图 10-1(a)所示。而此时 D 的鉴赏水平也才起步，只提出轮廓不够鲜明的意见。然后，G 开始针对如何绘制边界展开学习。

经过长时间努力，G 的作品已经可以看起来人物重点突出，如图 10-1(b)所示。而 D 这段时间通过反复观摩 K 的作品，鉴赏水平已有大幅提升，指出 G 在色彩搭配方面根本无法和 K 相比。

又经过了很久，G 不断总结吸收之前的经验教训而继续学习改进，日益颜色搭配合理，如图 10-1(c)所示。但此时 D 已经上升到了鉴赏家水平，再次指出批评意见，例如不够灵性，形似而神不似。如此这番重复多次，直到 G 能够随心所欲绘出如图 10-1(d)所示的作品，才让 D 无法分辨它和 K 作品之间的差别，即可成功地以假乱真。

G 的逻辑结构如图 10-2 所示，作品交给 D 进行判断的过程如图 10-3 所示。本质上可以用"图灵学习"对 GAN 进行概括，双方通过不断地重复对抗成长过程使得技术都日益精湛，直至达到某种均衡。

10.1.2 安装操作

本节所用代码文件结构如表 10-1 所示。

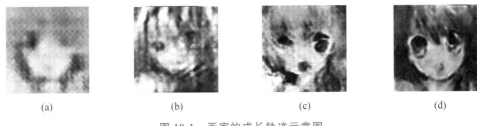

图 10-1 画家的成长轨迹示意图

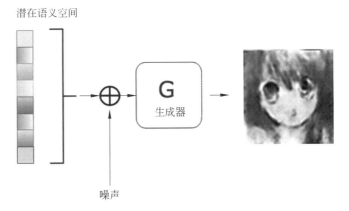

图 10-2 转置卷积构成的生成网络

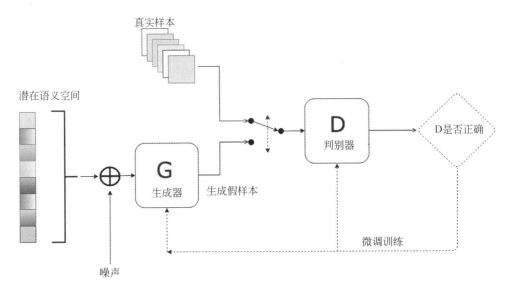

图 10-3 生成网络和判别网络

表 10-1 代码文件结构

文 件 名 称	实 现 功 能	文 件 名 称	实 现 功 能
dataset.py	处理数据集	gan_train.py	训练生成对抗网络
gan.py	测试生成对抗网络		

安装步骤如下。

(1) 启动 Anaconda Prompt 命令行,进入本章根目录。

(2) 进入本次实战目录,输入 python gan_train.py 进行训练。

(3) 训练完成后,输入 python gan.py 进行测试。

10.1.3 代码解析

本次实战核心代码主要包括 dataset.py、gan_train.py、gan.py。

下面的代码命名为 gan_train.py,主要功能是生成器生成图像,用来模仿漫画中的人物头像;判别器对这些任务头像进行判断,符合基本要求则保留,否则舍弃。通过上述的生成对抗网络,逐渐产生风格与样式愈发逼真的漫画人脸图像。

```
import   os
import   numpy as np
import   tensorflow as tf
from     tensorflow import keras
from     scipy.misc import toimage
import   glob
from     gan import Generator, Discriminator
from     dataset import make_anime_dataset
def save_result(val_out, val_block_size, image_path, color_mode):
    def preprocess(img):
        img = ((img + 1.0) * 127.5).astype(np.uint8)
        return img
    preprocesed = preprocess(val_out)
    final_image = np.array([])
    single_row = np.array([])
    for b in range(val_out.shape[0]):
        #拼接图像到一行上
        if single_row.size == 0:
            single_row = preprocesed[b, :, :, :]
        else:
            single_row = np.concatenate((single_row, preprocesed[b, :, :, :]), axis=1)
        #将一行的图像合并到最终图像上
        if (b+1) % val_block_size == 0:
            if final_image.size == 0:
                final_image = single_row
            else:
```

```python
                    final_image = np.concatenate((final_image, single_row), axis=0)
                #改变单行图像的尺寸
                single_row = np.array([])
        if final_image.shape[2] == 1:
            final_image = np.squeeze(final_image, axis=2)
        toimage(final_image).save(image_path)
def celoss_ones(logits):
        #计算判别结果与标签1的交叉熵
        y = tf.ones_like(logits)
        loss = keras.losses.binary_crossentropy(y, logits, from_logits=True)
        return tf.reduce_mean(loss)
def celoss_zeros(logits):
        #计算判别结果与标签0的交叉熵
        y = tf.zeros_like(logits)
        loss = keras.losses.binary_crossentropy(y, logits, from_logits=True)
        return tf.reduce_mean(loss)
def d_loss_fn(generator, discriminator, batch_z, batch_x, is_training):
        #计算判别器的损失函数
        #生成图像
        fake_image = generator(batch_z, is_training)
        #判定图像
        d_fake_logits = discriminator(fake_image, is_training)
        d_real_logits = discriminator(batch_x, is_training)
        #计算真实图像与1之间的误差,生成图像与0之间的误差,返回总体损失
        d_loss_real = celoss_ones(d_real_logits)
        d_loss_fake = celoss_zeros(d_fake_logits)
        loss = d_loss_fake + d_loss_real
        return loss
def g_loss_fn(generator, discriminator, batch_z, is_training):
        #计算生成器损失
        #采样生成图像
        fake_image = generator(batch_z, is_training)
        #在训练生成网络时,需要迫使生成图像判定为真
        d_fake_logits = discriminator(fake_image, is_training)
        #计算生成图像与1之间的误差
        loss = celoss_ones(d_fake_logits)
        return loss
def main():
    tf.random.set_seed(3333)
    np.random.seed(3333)
    os.environ['TF_CPP_MIN_LOG_LEVEL'] = '2'
    assert tf.__version__.startswith('2.')
    z_dim = 100
    epochs = 3000000                                            #训练步数
    batch_size = 64
```

```python
    learning_rate = 0.0002
    is_training = True
    #获取数据集路径
    img_path = glob.glob(r'C:\Users\z390\Downloads\anime-faces\*\*.jpg') + \
        glob.glob(r'C:\Users\z390\Downloads\anime-faces\*\*.png')
    print('images num:', len(img_path))
    #构建数据集对象
    dataset, img_shape, _ = make_anime_dataset(img_path, batch_size, resize=64)
    print(dataset, img_shape)
    sample = next(iter(dataset))                            #采样
    print(sample.shape, tf.reduce_max(sample).numpy(),
        tf.reduce_min(sample).numpy())
    dataset = dataset.repeat(100)                           #重复循环
    db_iter = iter(dataset)
    generator = Generator()                                 #创建生成器
    generator.build(input_shape = (4, z_dim))
    discriminator = Discriminator()                         #创建判别器
    discriminator.build(input_shape=(4, 64, 64, 3))
    #分别为生成器和判别器创建优化器
    g_optimizer = keras.optimizers.Adam(learning_rate=learning_rate, beta_1=0.5)
    d_optimizer = keras.optimizers.Adam(learning_rate=learning_rate, beta_1=0.5)
    generator.load_weights('generator.ckpt')
    discriminator.load_weights('discriminator.ckpt')
    print('Loaded chpt!!')
    d_losses, g_losses = [],[]
    for epoch in range(epochs):                             #训练 epochs 次
        #1. 训练判别器
        for _ in range(1):
            #采样隐藏向量
            batch_z = tf.random.normal([batch_size, z_dim])
            batch_x = next(db_iter)                         #采样真实图像
            #判别器前向计算
            with tf.GradientTape() as tape:
                d_loss = d_loss_fn(generator, discriminator, batch_z, batch_x, is_training)
            grads = tape.gradient(d_loss, discriminator.trainable_variables)
            d_optimizer.apply_gradients(zip(grads, discriminator.trainable_variables))
        #2. 训练生成器
        #采样隐藏向量
        batch_z = tf.random.normal([batch_size, z_dim])
        batch_x = next(db_iter)                             #采样真实图像
        #生成器前向计算
        with tf.GradientTape() as tape:
            g_loss = g_loss_fn(generator, discriminator, batch_z, is_training)
```

```python
            grads = tape.gradient(g_loss, generator.trainable_variables)
            g_optimizer.apply_gradients(zip(grads, generator.trainable_variables))
        if epoch % 100 == 0:
            print(epoch, 'd-loss:',float(d_loss), 'g-loss:', float(g_loss))
            #可视化
            z = tf.random.normal([100, z_dim])
            fake_image = generator(z, training=False)
            img_path = os.path.join('gan_images', 'gan-%d.png'%epoch)
            save_result(fake_image.numpy(), 10, img_path, color_mode='P')
            d_losses.append(float(d_loss))
            g_losses.append(float(g_loss))
            if epoch % 10000 == 1:
                #print(d_losses)
                #print(g_losses)
                generator.save_weights('generator.ckpt')
                discriminator.save_weights('discriminator.ckpt')
if __name__ == '__main__':
    main()
```

程序 gan.py 是生成器与判定器模型的程序实现。

```
import   tensorflow as tf
from     tensorflow import keras
from     tensorflow.keras import layers
class Generator(keras.Model):
    #生成器网络
    def __init__(self):
        super(Generator, self).__init__()
        filter = 64
        #转置卷积层 1,输出 channel 为 filter * 8,核大小为 4,步长为 1,不使用 padding,不
        #使用偏置,以下 4 个卷积层结构相似
        self.conv1 = layers.Conv2DTranspose(filter * 8, 4,1, 'valid', use_bias=False)
        self.bn1 = layers.BatchNormalization()
        self.conv2 = layers.Conv2DTranspose(filter * 4, 4,2, 'same', use_bias=False)
        self.bn2 = layers.BatchNormalization()
        self.conv3 = layers.Conv2DTranspose(filter * 2, 4,2, 'same', use_bias=False)
        self.bn3 = layers.BatchNormalization()
        self.conv4 = layers.Conv2DTranspose(filter * 1, 4,2, 'same', use_bias=False)
        self.bn4 = layers.BatchNormalization()
        self.conv5 = layers.Conv2DTranspose(3, 4,2, 'same', use_bias=False)
    def call(self, inputs, training=None):
        x = inputs #[z, 100]
        #Reshape 乘 4D 张量,方便后续转置卷积运算:(b, 1, 1, 100)
        x = tf.reshape(x, (x.shape[0], 1, 1, x.shape[1]))
```

```python
            x = tf.nn.relu(x)                                   #激活函数
            #转置卷积-BN-激活函数:(b, 4, 4, 512)
            x = tf.nn.relu(self.bn1(self.conv1(x), training=training))
            #转置卷积-BN-激活函数:(b, 8, 8, 256)
            x = tf.nn.relu(self.bn2(self.conv2(x), training=training))
            #转置卷积-BN-激活函数:(b, 16, 16, 128)
            x = tf.nn.relu(self.bn3(self.conv3(x), training=training))
            #转置卷积-BN-激活函数:(b, 32, 32, 64)
            x = tf.nn.relu(self.bn4(self.conv4(x), training=training))
            #转置卷积-激活函数:(b, 64, 64, 3)
            x = self.conv5(x)
            x = tf.tanh(x)                                      #输出x范围-1~1,与预处理一致
            return x
class Discriminator(keras.Model):
    #判别器
    def __init__(self):
        super(Discriminator, self).__init__()
        filter = 64
        #卷积层1,输出channel为filter,核大小为4,步长为2,不使用padding,不使用偏
#置,以下4个卷积层结构相似
        self.conv1 = layers.Conv2D(filter, 4, 2, 'valid', use_bias=False)
        self.bn1 = layers.BatchNormalization()
        self.conv2 = layers.Conv2D(filter*2, 4, 2, 'valid', use_bias=False)
        self.bn2 = layers.BatchNormalization()
        self.conv3 = layers.Conv2D(filter*4, 4, 2, 'valid', use_bias=False)
        self.bn3 = layers.BatchNormalization()
        self.conv4 = layers.Conv2D(filter*8, 3, 1, 'valid', use_bias=False)
        self.bn4 = layers.BatchNormalization()
        self.conv5 = layers.Conv2D(filter*16, 3, 1, 'valid', use_bias=False)
        self.bn5 = layers.BatchNormalization()
        #全局池化层
        self.pool = layers.GlobalAveragePooling2D()
        #特征打平
        self.flatten = layers.Flatten()
        #二分类全连接层,鉴别图像是真实图像还是虚假图像
        self.fc = layers.Dense(1)
    def call(self, inputs, training=None):
        #卷积-BN-激活函数:(4, 31, 31, 64)
        x = tf.nn.leaky_relu(self.bn1(self.conv1(inputs), training=training))
        #卷积-BN-激活函数:(4, 14, 14, 128)
        x = tf.nn.leaky_relu(self.bn2(self.conv2(x), training=training))
        #卷积-BN-激活函数:(4, 6, 6, 256)
        x = tf.nn.leaky_relu(self.bn3(self.conv3(x), training=training))
        #卷积-BN-激活函数:(4, 4, 4, 512)
        x = tf.nn.leaky_relu(self.bn4(self.conv4(x), training=training))
```

```
            #卷积-BN-激活函数:(4, 2, 2, 1024)
            x = tf.nn.leaky_relu(self.bn5(self.conv5(x), training=training))
            #卷积-BN-激活函数:(4, 1024)
            x = self.pool(x)
            #打平
            x = self.flatten(x)
            #输出,[b, 1024] => [b, 1]
            logits = self.fc(x)
            return logits
def main():
    d = Discriminator()
    g = Generator()
    x = tf.random.normal([2, 64, 64, 3])
    z = tf.random.normal([2, 100])
    prob = d(x)
    print(prob)
    x_hat = g(z)
    print(x_hat.shape)
if __name__ == '__main__':
    main()
```

10.1.4 训练测试

运行 python gan_train.py 进行训练,训练中使用的漫画人脸数据集如图 10-4 所示。

图 10-4 训练中使用的漫画人脸数据集

运行 python gan.py 生成新的卡通漫画人物图像,如图 10-5 所示。

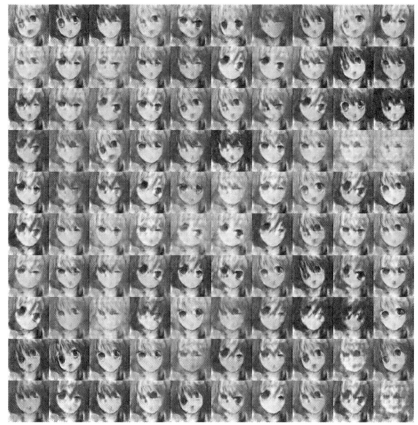

图 10-5　生成的一些卡通漫画人物图像

10.2　应用生成对抗网络 Pix2Pix 模仿欧式建筑风格

在本次实战中,将实现卷积 Pixel2Pixel 模型,并在 facades 数据集上对其进行训练。目的是通过训练将已有的基本建筑框架图像转换为欧式风格建筑照片。

10.2.1　背景原理

Pix2Pix 是一个图像翻译通用框架,它的作用是将一幅图像转换或者生成另外一幅图像。例如根据一幅素描画生成高清立体图像,或者将一幅高清立体图像生成一幅素描画。在图像处理过程中,如果要进行梯度图或彩色图之间的转换,则需要使用特定的算法来处理,这些算法的本质都是原像素到目的像素的映射转换。Pix2Pix 提供的是一种图像转换算法通用框架,能够统一解决所有图像像素之间的转换问题。

10.2.2　安装操作

本节所用代码文件结构如表 10-2 所示。

表 10-2　代码文件结构

文 件 名 称	实 现 功 能	文 件 名 称	实 现 功 能
gd.py	定义生成器与鉴别器	main.py	训练模型

安装步骤如下。

(1) 进入 Anaconda Prompt 命令行,切换至本章根目录。

(2) 进入本节代码的目录。

(3) 在命令行中输入 python main.py。

(4) 运行代码,在 images 文件夹下查看各轮次的运行结果。

10.2.3　代码解析

在之前的章节中已经对生成对抗网络进行了详细的介绍,因此以下仅对 main.py 文件进行解析。

(1) 导入必要的函数库,包括画图和图像处理的函数库 numpy、matplotlib 等。

```
import os
os.environ["CUDA_VISIBLE_DEVICES"]="-1"
import   tensorflow as tf
import   numpy as np


from    tensorflow import keras
import   time
from    matplotlib import pyplot as plt
from    gd import Discriminator, Generator
```

(2) 加载数据库并定义所需训练的图像大小统一定义为 256 像素×256 像素。

```
batch_size = 1
IMG_WIDTH = 256
IMG_HEIGHT = 256
path_to_zip = keras.utils.get_file('facades.tar.gz',
                        cache_subdir=os.path.abspath('.'),
origin = ' https://people. eecs. berkeley. edu/~ tinghuiz/projects/pix2pix/
datasets/facades.tar.gz',
                        extract=True)
PATH = os.path.join(os.path.dirname(path_to_zip), 'facades/')
def load_image(image_file, is_train):
    image = tf.io.read_file(image_file)
    image = tf.image.decode_jpeg(image)
    w = image.shape[1]
    w = w // 2
    real_image = image[:, :w, :]
    input_image = image[:, w:, :]
```

```python
        input_image = tf.cast(input_image, tf.float32)
        real_image = tf.cast(real_image, tf.float32)
        if is_train:
            #随机抖动
            #尺寸变为 286 x 286 x 3
            input_image = tf.image.resize(input_image, [286, 286])
            real_image = tf.image.resize(real_image, [286, 286])
            #对图像进行随机裁剪,裁剪后的图像尺寸为 256 x 256 x 3
            stacked_image = tf.stack([input_image, real_image], axis=0)
            cropped_image = tf.image.random_crop(stacked_image, size=[2, IMG_HEIGHT, IMG
_WIDTH, 3])
            input_image, real_image = cropped_image[0], cropped_image[1]
            if np.random.random() > 0.5:
                #随机翻转
                input_image = tf.image.flip_left_right(input_image)
                real_image = tf.image.flip_left_right(real_image)
        else:
            input_image = tf.image.resize(input_image, size=[IMG_HEIGHT, IMG_WIDTH])
            real_image = tf.image.resize(real_image, size=[IMG_HEIGHT, IMG_WIDTH])
        #将图像像素归一化到 [-1, 1]
        input_image = (input_image / 127.5) - 1
        real_image = (real_image / 127.5) - 1
        out = tf.concat([input_image, real_image], axis=2)
        return out
train_dataset = tf.data.Dataset.list_files(PATH+'/train/*.jpg')
train_iter = iter(train_dataset)
train_data = []
for x in train_iter:
    train_data.append(load_image(x, True))
train_data = tf.stack(train_data, axis=0)
#[800, 256, 256, 3]
print('train:', train_data.shape)
train_dataset = tf.data.Dataset.from_tensor_slices(train_data)
train_dataset = train_dataset.shuffle(400).batch(1)
test_dataset = tf.data.Dataset.list_files(PATH+'test/*.jpg')
test_iter = iter(test_dataset)
test_data = []
for x in test_iter:
    test_data.append(load_image(x, False))
test_data = tf.stack(test_data, axis=0)
#[800, 256, 256, 3]
print('test:', test_data.shape)
test_dataset = tf.data.Dataset.from_tensor_slices(test_data)
test_dataset = test_dataset.shuffle(400).batch(1)
generator = Generator()
```

```python
generator.build(input_shape=(batch_size, 256, 256, 3))
generator.summary()
discriminator = Discriminator()
discriminator.build(input_shape=[(batch_size, 256, 256, 3), (batch_size, 256, 256, 3)])
discriminator.summary()
g_optimizer = keras.optimizers.Adam(learning_rate=2e-4, beta_1=0.5)
d_optimizer = keras.optimizers.Adam(learning_rate=2e-4, beta_1=0.5)
```

（3）定义判别器和生成器的损失函数和生成函数。

```python
def discriminator_loss(disc_real_output, disc_generated_output):
    #计算判别器损失
    real_loss = keras.losses.binary_crossentropy(tf.ones_like(disc_real_output), disc_real_output, from_logits=True)
    generated_loss = keras.losses.binary_crossentropy(tf.zeros_like(disc_generated_output), disc_generated_output, from_logits=True)
    real_loss = tf.reduce_mean(real_loss)
    generated_loss = tf.reduce_mean(generated_loss)
    total_disc_loss = real_loss + generated_loss
    return total_disc_loss
def generator_loss(disc_generated_output, gen_output, target):
    #计算生成器损失
    LAMBDA = 100
    gan_loss = keras.losses.binary_crossentropy(tf.ones_like(disc_generated_output), disc_generated_output, from_logits=True)
    l1_loss = tf.reduce_mean(tf.abs(target - gen_output))
    gan_loss = tf.reduce_mean(gan_loss)
    total_gen_loss = gan_loss + (LAMBDA * l1_loss)
    return total_gen_loss
def generate_images(model, test_input, tar, epoch):
#生成并保存图像
    prediction = model(test_input, training=True)
    plt.figure(figsize=(15,15))
    display_list = [test_input[0], tar[0], prediction[0]]
    title = ['Input Image', 'Ground Truth', 'Predicted Image']
    for i in range(3):
        plt.subplot(1, 3, i+1)
        plt.title(title[i])
        plt.imshow(display_list[i] * 0.5 + 0.5)
        plt.axis('off')
    plt.savefig('images/epoch%d.png'%epoch)
    print('saved images.')
```

（4）调用上面所有的函数并运行主函数，从而得到训练模型。

```python
def main():
```

```python
    epochs = 100
    for epoch in range(epochs):
        start = time.time()
        for step, inputs in enumerate(train_dataset):
            input_image, target = tf.split(inputs, num_or_size_splits=[3, 3], axis=3)
            with tf.GradientTape() as gen_tape, tf.GradientTape() as disc_tape:
                #生成图像
                gen_output = generator(input_image, training=True)
                #鉴别图像
                disc_real_output = discriminator([input_image, target], training=True)
                #与原图对比计算损失
                disc_generated_output = discriminator([input_image, gen_output], training=True)
                gen_loss = generator_loss(disc_generated_output, gen_output, target)
                disc_loss = discriminator_loss(disc_real_output, disc_generated_output)
            generator_gradients = gen_tape.gradient(gen_loss, generator.trainable_variables)
            g_optimizer.apply_gradients(zip(generator_gradients, generator.trainable_variables))
            discriminator_gradients = disc_tape.gradient(disc_loss, discriminator.trainable_variables)
            d_optimizer.apply_gradients(zip(discriminator_gradients, discriminator.trainable_variables))
            if step% 100 == 0:
                #输出损失

                print(epoch, step, float(disc_loss), float(gen_loss))
        if epoch % 1 == 0:
            for inputs in test_dataset:
                input_image, target = tf.split(inputs, num_or_size_splits=[3, 3], axis=3)
                generate_images(generator, input_image, target, epoch)
                break
        print('Time taken for epoch {} is {} sec\n'.format(epoch + 1, time.time() - start))
    for inputs in test_dataset:
        input_image, target = tf.split(inputs, num_or_size_splits=[3, 3], axis=3)
        generate_images(generator, input_image, target, 99999)
        break
if __name__ == '__main__':
    main()
```

10.2.4 训练测试

运行 main.py，得到结果如图 10-6 所示。

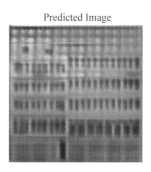
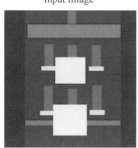

图 10-6　测试命令和一些测试结果

每行第一张图像是建筑物的结构图像，第二张图像是该图像对应的真实图像，第三张图像是使用训练好的模型生成的图像。第一行是训练的第一个轮次的效果，此时模型还不能很好地分辨建筑物的各个部分，生成的图像很模糊，对于屋顶和天空也不能进行明显区分。第二行是训练了 99 999 个轮次后模型的生成效果，生成的图像与真实图像已经非常接近了，模型能够正确区分并生成内容、环境、墙壁、窗户、阳台等各个部分区别十分明显。

10.3　应用循环对抗网络 CycleGAN 完成相似动物转换

在本次实战中，将实现 CycleGAN 并在 horse2zebra 数据集上训练完成斑马到马或者马到斑马的相似动物样式转换。

10.3.1　背景原理

Pix2Pix 在实际应用中有一个非常大的难题：训练数据需要原图像和目标图像成对出现。在实际的工业生产中，获取符合要求的训练数据的成本是比较高的。CycleGAN 是一种新的无监督图像前一通用框架，可以在没有成对训练数据的情况下将图像数据从源域迁移到目标域[45]。CycleGAN 的核心理念是转换互逆。举例来说，如果 F 和 G 是转换互逆的，那么 G 可以将 X 域的图像转换为 Y 域的风格，F 就可以将 Y 域的图像转换为 X 域的风格。这种转换的互逆性，就是 CycleGAN 应用于图像风格迁移领域的秘诀，例如将图像转换为抽象派风格等。

10.3.2 安装操作

本节所用代码文件结构如表 10-3 所示。

表 10-3 代码文件结构

文 件 名 称	实 现 功 能	文 件 名 称	实 现 功 能
model.py	定义编码器、解码器、生成器、判别器以及损失函数等	main.py	训练模型

安装步骤如下。

(1) 进入 Anaconda Prompt 命令行，切换至本章根目录。
(2) 进入本节代码的目录。
(3) 在命令行中输入 python main.py。
(4) 运行代码，在 images 文件夹下查看各轮次的运行结果。

10.3.3 代码解析

(1) 导入相关的函数库。

```
import  os
import  time
import  numpy as np
import  matplotlib.pyplot as plt
import  tensorflow as tf
import  numpy as np
from    tensorflow import keras
from    model import Generator, Discriminator, cycle_consistency_loss, generator
_loss, discriminator_loss
```

(2) 通常 GPU 不能动态分配内存，可以运用下面的语句来防止内存溢出。

```
physical_devices = tf.config.experimental.list_physical_devices('GPU')
assert len(physical_devices) > 0, "Not enough GPU hardware devices available"
tf.config.experimental.set_memory_growth(physical_devices[0], True)
tf.random.set_seed(22)
np.random.seed(22)
os.environ['TF_CPP_MIN_LOG_LEVEL'] = '2'
assert tf.__version__.startswith('2.')
learning_rate = 0.0002
batch_size = 1                          #Set batch size to 4 or 16 if training multigpu
img_size = 256
cyc_lambda = 10
epochs = 1000
```

(3) 下载数据集。

```
path_to_zip = keras.utils.get_file('horse2zebra.zip', cache_subdir=os.path.
```

```python
    abspath('.'), origin = ' https://people.eecs.berkeley.edu/~taesung_park/
CycleGAN/datasets/horse2zebra.zip', extract=True)
PATH = os.path.join(os.path.dirname(path_to_zip), 'horse2zebra/')
trainA_path = os.path.join(PATH, "trainA")
trainB_path = os.path.join(PATH, "trainB")
trainA_size = len(os.listdir(trainA_path))
trainB_size = len(os.listdir(trainB_path))
print('train A:', trainA_size)
print('train B:', trainB_size)
```

（4）载入数据及并通过 Shuffle 方式进行数据增广。

```python
def load_image(image_file):
    image = tf.io.read_file(image_file)
    image = tf.image.decode_jpeg(image, channels=3)
    #already 转换尺寸
    image = tf.image.convert_image_dtype(image, tf.float32)
    image = tf.image.resize(image, [256, 256])    image = image * 2 - 1
    return image
train_datasetA = tf.data.Dataset.list_files(PATH + 'trainA/*.jpg', shuffle=False)
train_datasetA = train_datasetA.shuffle(trainA_size).repeat(epochs)
train_datasetA = train_datasetA.map(lambda x: load_image(x))
train_datasetA = train_datasetA.batch(batch_size)
train_datasetA = train_datasetA.prefetch(batch_size)
train_datasetA = iter(train_datasetA)
train_datasetB = tf.data.Dataset.list_files(PATH + 'trainB/*.jpg', shuffle=False)
train_datasetB = train_datasetB.shuffle(trainB_size).repeat(epochs)
train_datasetB = train_datasetB.map(lambda x: load_image(x))
train_datasetB = train_datasetB.batch(batch_size)
train_datasetB = train_datasetB.prefetch(batch_size)
train_datasetB = iter(train_datasetB)
a = next(train_datasetA)
print('img shape:', a.shape, a.numpy().min(), a.numpy().max())
```

（5）定义生成器和判别器，并运用它们进行对抗训练。

```python
discA = Discriminator()
discB = Discriminator()
genA2B = Generator()
genB2A = Generator()
discA_optimizer = keras.optimizers.Adam(learning_rate, beta_1=0.5)
discB_optimizer = keras.optimizers.Adam(learning_rate, beta_1=0.5)
genA2B_optimizer = keras.optimizers.Adam(learning_rate, beta_1=0.5)
genB2A_optimizer = keras.optimizers.Adam(learning_rate, beta_1=0.5)
def generate_images(A, B, B2A, A2B, epoch):
    plt.figure(figsize=(15, 15))
```

```python
        A = tf.reshape(A, [256, 256, 3]).numpy()
        B = tf.reshape(B, [256, 256, 3]).numpy()
        B2A = tf.reshape(B2A, [256, 256, 3]).numpy()
        A2B = tf.reshape(A2B, [256, 256, 3]).numpy()
        display_list = [A, B, A2B, B2A]
        title = ['A', 'B', 'A2B', 'B2A']
        for i in range(4):
            plt.subplot(2, 2, i + 1)
            plt.title(title[i])
            #从[0,1]中获取像素并生成图像
            plt.imshow(display_list[i] * 0.5 + 0.5)
            plt.axis('off')
        plt.savefig('images/generated_%d.png'%epoch)
        plt.close()
def train(train_datasetA, train_datasetB, epochs, lsgan=True, cyc_lambda=10):
    for epoch in range(epochs):
        start = time.time()
        with tf.GradientTape() as genA2B_tape, tf.GradientTape() as genB2A_tape, \
                tf.GradientTape() as discA_tape, tf.GradientTape() as discB_tape:
            try:
                #下一个最小训练批次,默认大小为1
                trainA = next(train_datasetA)
                trainB = next(train_datasetB)
            except tf.errors.OutOfRangeError:
                print("Error, run out of data")
                break
            genA2B_output = genA2B(trainA, training=True)
            genB2A_output = genB2A(trainB, training=True)
            discA_real_output = discA(trainA, training=True)
            discB_real_output = discB(trainB, training=True)
            discA_fake_output = discA(genB2A_output, training=True)
            discB_fake_output = discB(genA2B_output, training=True)
            reconstructedA = genB2A(genA2B_output, training=True)
            reconstructedB = genA2B(genB2A_output, training=True)
            #生成图像
            #利用50个历史缓冲计算判别器损失值
            discA_loss = discriminator_loss(discA_real_output, discA_fake_output,
lsgan=lsgan)
            discB_loss = discriminator_loss(discB_real_output, discB_fake_output,
lsgan=lsgan)
            genA2B_loss = generator_loss(discB_fake_output, lsgan=lsgan) + \
                        cycle_consistency_loss(trainA, trainB, reconstructedA,
reconstructedB, cyc_lambda=cyc_lambda)
            genB2A_loss = generator_loss(discA_fake_output, lsgan=lsgan) + \
                        cycle_consistency_loss(trainA, trainB, reconstructedA,
reconstructedB, cyc_lambda=cyc_lambda)
        genA2B_gradients = genA2B_tape.gradient(genA2B_loss, genA2B.trainable_
```

```
variables)
        genB2A_gradients = genB2A_tape.gradient(genB2A_loss, genB2A.trainable_variables)
        discA_gradients = discA_tape.gradient(discA_loss, discA.trainable_variables)
        discB_gradients = discB_tape.gradient(discB_loss, discB.trainable_variables)
        genA2B_optimizer.apply_gradients(zip(genA2B_gradients, genA2B.trainable_variables))
        genB2A_optimizer.apply_gradients(zip(genB2A_gradients, genB2A.trainable_variables))
        discA_optimizer.apply_gradients(zip(discA_gradients, discA.trainable_variables))
        discB_optimizer.apply_gradients(zip(discB_gradients, discB.trainable_variables))
        if epoch % 40 == 0:
            generate_images(trainA, trainB, genB2A_output, genA2B_output, epoch)
            print('Time taken for epoch {} is {} sec'.format(epoch + 1, time.time() - start))
if __name__ == '__main__':
    train(train_datasetA, train_datasetB, epochs=epochs, lsgan=True, cyc_lambda=cyc_lambda)
```

10.3.4 训练测试

运行 main.py，得到训练的每一轮次的图像替换图像如图 10-7 所示。

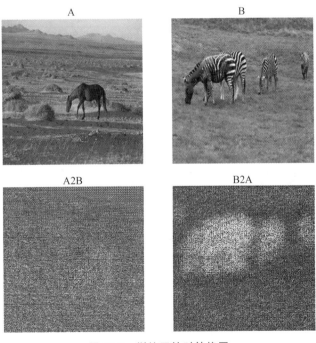

图 10-7　训练开始时转换图

图 10-7 是训练刚开始运行到 41 轮时模型测试效果,可以看到动物风格转换并不明显,运行到 1000 轮次以上时,可以看到图 10-8 中斑马和马的转换效果就非常明显了。

图 10-8 训练结果

10.4 WGAN-GP 人脸生成实战

本次实战将使用 WGAN-GP 进行人脸生成[53]。WGAN-GP 的不同之处在于引入 Wasserteion 距离来衡量两个样本在高维空间中的距离,同时在判别器的损失中加入了一个正则项,即梯度惩罚(Gradient Penalty,GP)。

10.4.1 背景原理

自从 2014 年 IanGoodfellow 等发表了生成对抗网络的论文之后,GAN 引起了工业界的广泛关注,关于 GAN 的越来越多的理论和实践被人们研究出来。目前 GAN 的大部分应用都与图像有关,可以用来生成逼真的图像。例如 StackGAN,以一段文本作为输入,输出一幅图像,这幅高清图像的内容与文本文字描述高度相关。再如 CycleGAN,可以将输入的一幅图像转换成另一个风格的图像。

理论上,GAN 可以生成任何想要生成的东西,当然前提是要设计并训练好模型。例如可以尝试用 GAN 来生成语音或者文本。再如,既然可以将马转换成斑马,那么是不是可以将图像中任何区域变成其他图像,如人脸的转换,将指定的人脸转换成其他人脸;衣服的转换,将西服变成 T 恤;情绪表情的转换,将严肃的表情变成笑脸。可以说,GAN 是一个非常强大的工具,其更多应用场景还需要靠大家的灵感和想象力。

上面说过 GAN 由 G 和 D 组成,可以使用任何方式实现 G 和 D。假如任务是让 G 学会生成真实的彩色图像，如人脸，那么就需要使用大量的人脸数据让 GAN 去学习。让网络能够看懂图像最好的方式就是使用卷积网络，这样使用卷积网络实现的 G 和 D 就叫作GCGAN。如果试着训练 GCGAN 生成图像，则会发现若不在一段时间内打印出生成器生成的图像，就不知道当前网络训练到什么程度，生成器和判别器的损失也无法指示当前网络的训练情况。另外，在网上可以搜到很多训练 GAN 的 trick，这是因为原始 GAN 的设计问题会导致训练不稳定、梯度消失、模式崩溃等，所以训练好 GAN 是比较困难的。

WGAN 的出现就是为了解决 GAN 的设计问题的。总之，判别器既不能训练得太好，又不能训练得太差。如果判别器训练得太好，则生成器的梯度会消失。因为生成器的梯度是靠判别器提供的，如果判别器达到了最优，对于真实图像给出的概率是常数 1，生成图像给出的概率是常数 0，这样生成器不会得到任何梯度，造成梯度消失。而如果判别器训练得不好，则生成器的梯度会不准确，导致生成器不收敛，也就学不到要生成图像的特征，即生成器不能生成真实的样本。例如判别器还没有训练好，此时如果生成器生成的假图像并不是很真实，一旦判别其给出了认为是真实图像的预测，就会导致生成器认为自己生成的图像是正确的，那么生成器就会按照这个假图像的模型继续生成，然后判别器仍然会给出正确的评价。结果就是生成器永远不会学到生成真实图像的特征，判别器和生成器都不会训练成功。或者生成器能够生成一些看起来较真实的图像，然后又尝试生成风格不太一样但也属于真实图像，此时如果判别器没有训练好的话，就会将该图像预测成假图像，并将惩罚梯度输入给生成器。这样生成器为了避免惩罚，则不会再尝试生成多样性的图像了，而是不断生成重复的图像，这就导致了模式崩溃。以上就是 GAN 不稳定的直观解释。

于是 WGAN 引入了 Wasserstein 距离这个概念，其优点在于，即使两个样本在高维空间中没有重叠，也可以衡量两者间的距离，并且可以提供有效梯度。但是使用 Wasserteion 距离需要满足连续性条件，称为 Lipschitz 连续。

WGAN-GP 的思想是在判别器的损失上加入一个正则项，这个正则项就是梯度惩罚（Gradient Penalty，GP）。

10.4.2 安装操作

本节所用代码文件结构如表 10-4 所示。

表 10-4 代码文件结构

文件名称	实现功能
face_generation-wgan_gp.ipynb	给图像做标题的核心文件

本次实战只有 face_generation-wgan_gp.ipynb 这一个代码文件，里面包含所有的图像增广操作。

安装步骤如下。

（1）进入 Anaconda Prompt 命令行，切换至本章根目录。

（2）在命令行中输入 jupyter notebook。

（3）在浏览器中打开 face_generation-wgan_gp.ipynb。

（4）分步运行代码，查看运行结果。

10.4.3 代码解析

本次实战核心代码为 face_generation-wgan_gp_tf2.ipynb。

```python
import tensorflow as tf
%matplotlib inline
import os
from glob import glob
from matplotlib import pyplot
physical_devices = tf.config.experimental.list_physical_devices('GPU')
assert len(physical_devices) > 0, 'Not enough GPU hardware devices available'
tf.config.experimental.set_memory_growth(physical_devices[0], True)
#数据集本地目录
data_dir = './data'
```

1. 辨别器

```python
OUTPUT_DIM = 28 * 28 * 3
def discriminator():
    model = tf.keras.Sequential([
        tf.keras.layers.Reshape([28, 28, 3], input_shape=([OUTPUT_DIM,])),
        tf.keras.layers.Conv2D(64, (5, 5), strides=(2, 2), padding='same'),
        tf.keras.layers.LeakyReLU(),
#        tf.keras.layers.Dropout(0.3),
        tf.keras.layers.Conv2D(128, (5, 5), strides=(2, 2), padding='same'),
        tf.keras.layers.LeakyReLU(),
#        tf.keras.layers.Dropout(0.3),
        tf.keras.layers.Conv2D(256, (5, 5), strides=(2, 2), padding='same'),
        tf.keras.layers.LeakyReLU(),
#        tf.keras.layers.Dropout(0.3),
        tf.keras.layers.Flatten(),
        tf.keras.layers.Dense(1)
        ])
    return model
```

2. 生成器

```python
z_dim = 128
def generator():
    model = tf.keras.Sequential([
        tf.keras.layers.Dense(2 * 2 * 512, activation="relu", input_shape=([z_dim,])),
        tf.keras.layers.Reshape([2, 2, 512]),
        tf.keras.layers.Conv2DTranspose(256, (5, 5), strides=(2, 2), activation="relu", padding='same'),
```

```
        tf.keras.layers.Conv2DTranspose(128, (5, 5), strides=(2, 2), activation
="relu", padding='same'),
        tf.keras.layers.Lambda(lambda x:x[:, :7, :7, :]),
        tf.keras.layers.Conv2DTranspose(64, (5, 5), strides=(2, 2), activation="
relu", padding='same'),
        tf.keras.layers.Conv2DTranspose(3, (5, 5), strides=(2, 2), activation="
tanh", padding='same'),
        tf.keras.layers.Reshape([OUTPUT_DIM])
        ])
    return model
```

3. 损失函数

```
def generator_loss(d_logits_fake):
    gen_cost = -tf.reduce_mean(d_logits_fake)
    return gen_cost
batch_size = 64
def discriminator_loss(input_real, g_output, d_logits_real, d_logits_fake):
    LAMBDA=10
    disc_cost = tf.reduce_mean(d_logits_fake) - tf.reduce_mean(d_logits_real)
    alpha = tf.random.uniform(
        shape=[batch_size, 1],
        minval=0.,
        maxval=1.
    )
    differences = g_output - input_real
    interpolates = input_real + (alpha * differences)
    gradients = tf.gradients(discriminator_model(interpolates, training=True),
[interpolates])[0]
    slopes = tf.sqrt(tf.reduce_sum(tf.square(gradients), axis=[1]))
#reduction_indices
    gradient_penalty = tf.reduce_mean((slopes - 1.) ** 2)
    disc_cost += LAMBDA * gradient_penalty
    return disc_cost
```

4. 优化器

```
beta1 = 0   #0.5
beta2 = 0.9
generator_optimizer = tf.keras.optimizers.Adam(1e-4, beta1, beta2)
discriminator_optimizer = tf.keras.optimizers.Adam(1e-4, beta1, beta2)
generator_model = generator()
discriminator_model = discriminator()
```

5. 训练神经网络

```
import math
```

```python
def images_square_grid(images, mode):
    #获取图像正方形网格的最大尺寸
    save_size = math.floor(np.sqrt(images.shape[0]))
    #缩放到 0~255
    images = (((images - images.min()) * 255) / (images.max() - images.min())).astype(np.uint8)
    #把图像按正方形排列
    images_in_square = np.reshape(
            images[:save_size * save_size],
            (save_size, save_size, images.shape[1], images.shape[2], images.shape[3]))
    if mode == 'L':
        images_in_square = np.squeeze(images_in_square, 4)
    #把图像合并到网格图像
    new_im = Image.new(mode, (images.shape[1] * save_size, images.shape[2] * save_size))
    for col_i, col_images in enumerate(images_in_square):
        for image_i, image in enumerate(col_images):
            im = Image.fromarray(image, mode)
            new_im.paste(im, (col_i * images.shape[1], image_i * images.shape[2]))
    return new_im
def show_generator_output(n_images, image_mode):
    cmap = None if image_mode == 'RGB' else 'gray'
    noise = tf.random.normal([n_images, z_dim])
    samples = generator_model(noise, training=False)
    samples=tf.reshape(samples, (n_images, 28, 28, 3))
    images_grid = images_square_grid(samples.numpy(), image_mode)
    pyplot.imshow(images_grid, cmap=cmap)
    pyplot.show()
@tf.function
def d_train_step(input_real):
        #从一个正态分布中生成噪声
    noise = tf.random.normal([batch_size, z_dim])
    with tf.GradientTape() as disc_tape:
        g_output = generator_model(noise, training=True)
        d_logits_real = discriminator_model(input_real, training=True)
        d_logits_fake = discriminator_model(g_output, training=True)
        disc_loss = discriminator_loss(input_real, g_output, d_logits_real, d_logits_fake)
    gradients_of_discriminator=disc_tape.gradient(disc_loss, discriminator_model.variables)    discriminator_optimizer.apply_gradients(zip(gradients_of_discriminator, discriminator_model.variables))
    return disc_loss
@tf.function
def g_train_step():
        #从一个正态分布中生成噪声
```

```python
        noise = tf.random.normal([batch_size, z_dim])
        with tf.GradientTape() as gen_tape:
            g_output = generator_model(noise, training=True)
            d_logits_fake = discriminator_model(g_output, training=True)
            gen_loss = generator_loss(d_logits_fake)
        gradients_of_generator = gen_tape.gradient(gen_loss, generator_model.variables)

        generator_optimizer.apply_gradients(zip(gradients_of_generator, generator_model.variables))
        return gen_loss
```

6. 训练函数

```python
import time
import numpy as np
from tensorflow.python.ops import summary_ops_v2
def train(epochs, batch_size, learning_rate, beta1, beta2, get_batches, Xs, data_image_mode,
          CRITIC_ITERS=5):
    train_times = 0
    gen = get_batches(Xs, batch_size)
    if True:
        timestamp = str(int(time.time()))
        save_dir = "./models"
        if tf.io.gfile.exists(save_dir):
            #            print('Removing existing model dir: {}'.format(MODEL_DIR))
            #            tf.io.gfile.rmtree(MODEL_DIR)
            pass
        else:
            tf.io.gfile.makedirs(save_dir)
        train_dir = os.path.join(save_dir, 'summaries', 'train')
        train_summary_writer = summary_ops_v2.create_file_writer(train_dir, flush_millis=10000)
        checkpoint_dir = os.path.join(save_dir, 'checkpoints')
        checkpoint_prefix = os.path.join(checkpoint_dir, 'ckpt')
        checkpoint = tf.train.Checkpoint(generator_optimizer=generator_optimizer,
            discriminator_optimizer=discriminator_optimizer, generator=generator_model,
            discriminator=discriminator_model)
        #如果存在检查点,则在创建时还原变量
        checkpoint.restore(tf.train.latest_checkpoint(checkpoint_dir))
        batch_step = 0
        avg_g_loss = tf.keras.metrics.Mean('g_loss', dtype=tf.float32)
        avg_d_loss = tf.keras.metrics.Mean('d_loss', dtype=tf.float32)
#        try:
```

```python
        if True:
            for epoch_i in range(epochs):
                if True:
                    start_time = time.time()
                    if epoch_i > 0:
                        gen_loss = g_train_step()
                    avg_d_loss.reset_states()
                    for i in range(CRITIC_ITERS):
                        batch_images = next(gen)
                        batch_images *= 2
                        batch_images=batch_images.reshape([-1, OUTPUT_DIM])
                        disc_loss = d_train_step(batch_images)
                        avg_d_loss(disc_loss)
#                         summary_ops_v2.scalar('disc_loss', disc_loss)
                    batch_step += 1
                    train_times += 1
                    if train_times % 10 == 0:
                        print(
                            "Epochs {}/{} Batch Step {}/{} gen_loss = {}..., disc_loss = {}..., train using time = {:.3f}".format(
                                epoch_i + 1, epochs, batch_step, len(Xs) // batch_size + 1, gen_loss,
                                avg_d_loss.result(), time.time() - start_time))
                    if train_times % 100 == 0:
                        checkpoint.save(checkpoint_prefix)
#                         print("Model saved in file: {}".format(save_path))
                        show_generator_output(16, data_image_mode)
                    if batch_step % (len(Xs) // batch_size) == 0:
                        batch_step = 0
#         except Exception as reason:
#             print("except!!!", type(reason), reason)
#             save_path = saver.save(sess, os.path.join(save_dir, 'best_model.ckpt'), global_step=step)
#             print("Model saved in file: {}".format(save_path))
    print("Done!")
from PIL import Image
IMAGE_WIDTH = 28
IMAGE_HEIGHT = 28
def get_batches(Xs, batch_size):
    while(True):
        for start in range(0, len(Xs), batch_size):
            end = min(start + batch_size, len(Xs))
            files=[]
            for file in Xs[start:end]:
                image = Image.open(file)
```

```
                j = (image.size[0] - 110) // 2
                i = (image.size[1] - 110) // 2
                image = image.crop([j, i, j + 110, i + 110])
                image = image.resize([IMAGE_WIDTH, IMAGE_HEIGHT], Image.BILINEAR)
                image = np.array(image)
                files.append(image)
            files=np.array(files).astype(np.float32)
            if len(files.shape) < 4:
                files=np.expand_dims(files, 0)
            yield files / 255 - 0.5
```

7. 开始训练

```
OUTPUT_DIM = 28 * 28 * 3
batch_size = 64  #128 #16 #64 128
z_dim = 128
learning_rate = 1e-4  #0.0003 #  1e-4
beta1 = 0  #0.5
beta2 = 0.9
epochs = 60000 #5 *
Xs = glob(os.path.join(data_dir, 'img_align_celeba/*.jpg'))
train(epochs, batch_size, learning_rate, beta1, beta2, get_batches, Xs, "RGB")
```

8. 批量生成人脸，参数是生成图像个数

```
class network(object):
    def __init__(self, epochs):
        self.epochs = epochs
        save_dir = "./models"
        if tf.io.gfile.exists(save_dir):
            #            print('Removing existing model dir: {}'.format(MODEL_DIR))
            #            tf.io.gfile.rmtree(MODEL_DIR)
            pass
        else:
            tf.io.gfile.makedirs(save_dir)

        beta1 = 0  #0.5
        beta2 = 0.9
        generator_optimizer = tf.keras.optimizers.Adam(1e-4, beta1, beta2)
        discriminator_optimizer = tf.keras.optimizers.Adam(1e-4, beta1, beta2)
        self.generator_model = generator()
        discriminator_model = discriminator()
        checkpoint_dir = os.path.join(save_dir, 'checkpoints')
        checkpoint_prefix = os.path.join(checkpoint_dir, 'ckpt')
```

```
        checkpoint = tf.train.Checkpoint(generator_optimizer=generator_optimizer,
discriminator_optimizer = discriminator_optimizer, generator = self.generator_
model, discriminator=discriminator_model)
        #如果存在检查点,则在创建时还原变量

        checkpoint.restore(tf.train.latest_checkpoint(checkpoint_dir))
        for epoch_i in range(self.epochs):
            show_generator_output(16, "RGB")
obj=network(3)
```

10.4.4 训练测试

运行 face_generation-wgan_gp_tf2.ipynb 这个程序,得到图 10-9 所示的图像。

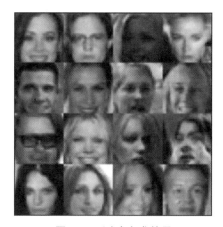

图 10-9 测试生成效果

第 11 章

强化学习与迁移学习

本章介绍了多种关于强化学习与迁移学习的背景知识,通过机器玩 *Flappy Bird* 和程序绘制油画等实战项目来加深关于强化学习与迁移学习的理论知识。

11.1 强化学习之玩转 *Flappy Bird*

本次实战主要通过小鸟自动飞跃障碍的游戏 *Flappy Bird*,来让大家对深度问答网络有一个直观的认识,充分了解强化学习的魅力[53]。

11.1.1 背景原理

小鸟自动飞跃障碍的游戏主要采用了深度 Q 网络。它就是一个卷积神经网络,用强化学习的变体进行训练,其输入是原始像素,其输出是估计未来回报的值函数。

深度 Q 网络将深度学习和强化学习结合在了一起。强化学习从奖励中进行学习,而这些奖励往往是稀疏的、有噪声、有延迟,并且状态序列高度相关[48]。为了克服这些问题,深度 Q 网络使用卷积神经网络从原始的视频数据中学习控制策略,并且引入了经验回放机制空间,从先前的状态转移中随机采样,使训练平稳进行[49]。

深度 Q 网络算法的伪代码如下所示。

```
Initialize replay memory D to size N
Initialize action-value function Q with random weights
for episode = 1, M do
    Initialize state s_1
    for t = 1, T do
        With probability ? select random action a_t
        otherwise select a_t=max_a  Q(s_t,a; θ_i)
        Execute action a_t in emulator and observe r_t and s_(t+1)
        Store transition (s_t,a_t,r_t,s_(t+1)) in D
        Sample a minibatch of transitions (s_j,a_j,r_j,s_(j+1)) from D
        Set y_j:=
            r_j for terminal s_(j+1)
            r_j+γ* max_(a^')   Q(s_(j+1),a'; θ_i) for non-terminal s_(j+1)
        Perform a gradient step on (y_j-Q(s_j,a_j; θ_i))^2 with respect to θ
    end for
end for
```

11.1.2 安装操作

本节所用代码文件结构如表 11-1 所示。

表 11-1 代码文件结构

文 件 名 称	实 现 功 能	文 件 名 称	实 现 功 能
flappy_bird_utils.py	*Flappy Bird* 游戏规则	wrapped_flappy_bird.py	游戏效果
deep_q_network.py	深度 Q 网络实现	deep_q_network_tf2.py	深度 Q 网络 tf2 实现

安装步骤如下。
(1) 进入 Anaconda Prompt 命令行,切换至本章根目录。
(2) 进入本节代码的目录。
(3) 在命令行中输入 deep_q_network_tf2.py。
(4) 运行代码,查看运行结果。

11.1.3 代码解析

本次实战核心代码为 python deep_q_network_tf2.py。下面进行分段讲解。
(1) 导入所必需的函数库和动态内存分配方法。

```
#!/usr/bin/env python
from __future__ import print_function
import tensorflow as tf
import cv2
import sys
sys.path.append("game/")
import wrapped_flappy_bird as game
import random
import numpy as np
from collections import deque
import os
from tensorflow.python.ops import summary_ops_v2
physical_devices = tf.config.experimental.list_physical_devices('GPU')
assert len(physical_devices) > 0, "Not enough GPU hardware devices available"
tf.config.experimental.set_memory_growth(physical_devices[0], True)
```

(2) 定义相关的训练参数。

```
GAME = 'bird' # the name of the game being played for log files
ACTIONS = 2 # number of valid actions
GAMMA = 0.99 # decay rate of past observations
OBSERVE = 100000. # timesteps to observe before training
EXPLORE = 2000000. # frames over which to anneal epsilon
FINAL_EPSILON = 0.0001 # final value of epsilon
INITIAL_EPSILON = 0.0001 # starting value of epsilon
```

```python
REPLAY_MEMORY = 50000 # number of previous transitions to remember
BATCH = 32 # size of minibatch
FRAME_PER_ACTION = 1
```

（3）创建深度问答网络。

```python
def createNetwork():
    # 输入层
    s = tf.keras.layers.Input(shape=(80, 80, 4), dtype='float32')   # tf.placeholder("float", [None, 80, 80, 4])
    # 隐藏层
    h_conv1 = tf.keras.layers.Conv2D(filters=32, kernel_size=8, strides=4, padding='same', activation="relu")(s)
    h_pool1 = tf.keras.layers.MaxPool2D(strides=2, padding='same')(h_conv1)
    h_conv2 = tf.keras.layers.Conv2D(filters=64, kernel_size=4, strides=2, padding='same', activation="relu")(h_pool1)
    h_conv3 = tf.keras.layers.Conv2D(filters=64, kernel_size=3, strides=1, padding='same', activation="relu")(h_conv2)
    h_conv3_flat = tf.keras.layers.Flatten()(h_conv3)
    h_fc1 = tf.keras.layers.Dense(units=512, activation='relu')(h_conv3_flat)
    # 读出层
    readout = tf.keras.layers.Dense(units=ACTIONS)(h_fc1)
    model = tf.keras.Model(
        inputs=[s],
        outputs=[readout])
    model.summary()
    return model
import pickle
def compute_loss(readout, a, y):
    readout_action = tf.reduce_sum(tf.multiply(readout, a), axis=1)
    cost = tf.reduce_mean(tf.square(tf.convert_to_tensor(y, dtype="float32") - readout_action))
    return cost
@tf.function
def train_step(model, minibatch, optimizer):
    # 记录用于计算损失的操作,以便于计算损失相对于变量的梯度
    with tf.GradientTape() as tape:
        # 如果值为真,则获取 batch 值
        s_j_batch = [d[0] for d in minibatch]
        a_batch = [d[1] for d in minibatch]
        r_batch = [d[2] for d in minibatch]
        s_j1_batch = [d[3] for d in minibatch]
        y_batch = []
        s_j1_batch = tf.stack(s_j1_batch)
        readout_j1_batch = model(s_j1_batch, training=True)
        for i in range(0, len(minibatch)):
```

```python
            terminal = minibatch[i][4]
            if terminal:
                y_batch.append(r_batch[i])
            else:
                y_batch.append(r_batch[i] + GAMMA * tf.keras.backend.max(readout_j1_batch[i]))
        #执行梯度步骤
        s_j_batch = tf.stack(s_j_batch)
        readout = model(s_j_batch, training=True)
        loss = compute_loss(readout, a_batch, y_batch)
    grads = tape.gradient(loss, model.trainable_variables)
    optimizer.apply_gradients(zip(grads, model.trainable_variables))
    return loss
```

（4）定义训练的网络模型与文件保存的具体位置。

```python
def trainNetwork(model):
    optimizer = tf.keras.optimizers.Adam(1e-6)  #tf.train.AdamOptimizer(1e-6).minimize(cost)
    MODEL_DIR = "./saved_networks"
    if tf.io.gfile.exists(MODEL_DIR):
        #            print('Removing existing model dir: {}'.format(MODEL_DIR))
        #            tf.io.gfile.rmtree(MODEL_DIR)
        pass
    else:
        tf.io.gfile.makedirs(MODEL_DIR)
    train_dir = os.path.join(MODEL_DIR, 'summaries', 'train')
    test_dir = os.path.join(MODEL_DIR, 'summaries', 'eval')
    train_summary_writer = summary_ops_v2.create_file_writer(train_dir, flush_millis=10000)
    test_summary_writer = summary_ops_v2.create_file_writer(test_dir, flush_millis=10000, name='test')
    checkpoint_dir = os.path.join(MODEL_DIR, 'checkpoints')
    checkpoint_prefix = os.path.join(checkpoint_dir, 'ckpt')
    checkpoint = tf.train.Checkpoint(model=model, optimizer=optimizer)
    #如果存在检查点,则在创建时还原变量
    checkpoint.restore(tf.train.latest_checkpoint(checkpoint_dir))
    #打开游戏状态与模拟器通信
    game_state = game.GameState()
    #将以前的观察结果存储在回放内存中
    D = deque()
    #printing
    a_file = open("logs_" + GAME + "/readout.txt", 'w')
    h_file = open("logs_" + GAME + "/hidden.txt", 'w')
    #通过不做任何事情获得第一个状态,并将图像预处理到 $80 \times 80 \times 4$
    do_nothing = np.zeros(ACTIONS)
```

```python
    do_nothing[0] = 1
    x_t, r_0, terminal = game_state.frame_step(do_nothing)
    x_t = cv2.cvtColor(cv2.resize(x_t, (80, 80)), cv2.COLOR_BGR2GRAY)
    #x_t = np.expand_dims(x_t, 2)
    #print(x_t.shape)
    ret, x_t = cv2.threshold(x_t, 1, 255, cv2.THRESH_BINARY)
    s_t = np.stack((x_t, x_t, x_t, x_t), axis=2)
    #print(s_t.shape)
    s_t = np.expand_dims(s_t, 0)
    #print(s_t.shape)
#开始训练
```

(5) 开始训练并通过分支语句定义好不同情境下面小鸟所做出的不同决策。

```python
epsilon = INITIAL_EPSILON
t = 0
while "flappy bird" != "angry bird":
    #贪心法缓则第一个动作的 epsilon
    readout_t = model(s_t.astype(np.float32), training=True)
    #print("readout_t.shape ", readout_t.shape)
    a_t = np.zeros([ACTIONS])
    action_index = 0
    if t % FRAME_PER_ACTION == 0:
        if random.random() <= epsilon:
            print("----------Random Action----------")
            action_index = random.randrange(ACTIONS)
            a_t[random.randrange(ACTIONS)] = 1
        else:
            action_index = np.argmax(readout_t)
            a_t[action_index] = 1
    else:
        a_t[0] = 1 #do nothing
    #缩小 epsilon
    if epsilon > FINAL_EPSILON and t > OBSERVE:
        epsilon -= (INITIAL_EPSILON - FINAL_EPSILON) / EXPLORE
    #执行查询到的动作,观察下一个状态并记录
    x_t1_colored, r_t, terminal = game_state.frame_step(a_t)
    x_t1 = cv2.cvtColor(cv2.resize(x_t1_colored, (80, 80)), cv2.COLOR_BGR2GRAY)
    ret, x_t1 = cv2.threshold(x_t1, 1, 255, cv2.THRESH_BINARY)
    x_t1 = np.reshape(x_t1, (1, 80, 80, 1))
    #s_t1 = np.append(x_t1, s_t[:,:,1:], axis = 2)
    #print("s_t.shape ", s_t.shape)    # (1, 80, 80, 4)
    #print("x_t1.shape ", x_t1.shape)   # (1, 80, 80, 1)
    s_t1 = np.append(x_t1, s_t[:, :, :, :3], axis=3)
    #print("s_t1.shape", s_t1.shape)    # (1, 80, 80, 4)
    #将转换存储在 D 中
```

```python
        D.append((s_t[0].astype(np.float32), a_t, r_t, s_t1[0].astype(np.float32),
terminal))
        if len(D) > REPLAY_MEMORY:
            D.popleft()
        #只有完成观察才能训练
        if t > OBSERVE:
            #取样一个小批量进行训练
            minibatch = random.sample(D, BATCH)
            train_step(model, minibatch, optimizer)
        #更新旧值
        s_t = s_t1
        t += 1
        #每 10 000 次迭代保存一次过程
        if t % 10000 == 0:
            checkpoint.save(checkpoint_prefix)
        #打印信息
        state = ""
        if t <= OBSERVE:
            state = "observe"
        elif t > OBSERVE and t <= OBSERVE + EXPLORE:
            state = "explore"
        else:
            state = "train"
        print("TIMESTEP", t, "/ STATE", state, \
            "/ EPSILON", epsilon, "/ ACTION", action_index, "/ REWARD", r_t, \
            "/ Q_MAX %e" % np.max(readout_t))
        #写入文件
        '''
        if t % 10000 <= 100:
            a_file.write(",".join([str(x) for x in readout_t]) + '\n')
            h_file.write(",".join([str(x) for x in h_fc1.eval(feed_dict={s:[s_t]})
[0]]) + '\n')
            cv2.imwrite("logs_tetris/frame" + str(t) + ".png", x_t1)
        '''
```

（6）最终的 main 函数调用上述所有的函数，进行程序主体的训练与效果展示。

```python
def playGame():
    model = createNetwork()
    trainNetwork(model)
def main():
    playGame()
if __name__ == "__main__":
    main()
```

11.1.4 训练测试

运行 python deep_q_network_tf2.py 得到小鸟的训练过程和实际的飞越障碍界面效果,如图 11-1 所示。

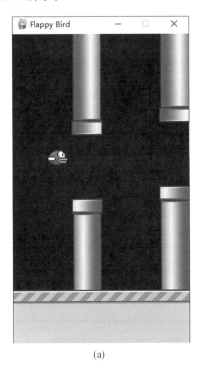
(a)

(b)

图 11-1 飞越障碍界面效果

11.2 使用 TensorFlow Hub 实现迁移学习预测影评分类

本次实战使用 TensorFlow Hub 实现迁移学习预测影评分类,评论文本将影评分为积极(Positive)或消极(Negative)两类。这是一个二元(Binary)或者二分类问题,一种重要且应用广泛的机器学习问题。在实战中主要采用的影评数据集为 IMDB 数据集,其包含 50 000 条影评文本。从该数据集切割出的 25 000 条评论用作训练,另外 25 000 条用作测试。训练集与测试集是平衡的(Balanced),意味着它们包含相等数量的积极和消极评论。

11.2.1 背景原理

随着越来越多的机器学习应用场景出现,而要得到一个好的模型需要大量标注好的数据[50],这一任务花费巨大且枯燥无味。因此,人们希望能够利用之前标注好的数据(一般是相关领域中的数据),从中习得知识或模式,并应用到相关领域的问题中。

一般来说,迁移学习通过使用已经训练好的模型,再加上一个很小的目标领域数据集,就可以将已有的经验和模式以一种高效的方式分享给新领域内的问题,而不必让模型从零开始学起。

11.2.2 安装操作

本节所用代码文件结构如表 11-2 所示。

表 11-2 代码文件结构

文 件 名 称	实 现 功 能
text_classification_with_hub.ipynb	预测图像种类

本次实战只有 text_classification_with_hub.ipynb 这一个代码文件,在这个代码中完成对图像种类的预测。

安装步骤如下。

(1)进入 Anaconda Prompt 命令行,切换至本章根目录。
(2)在命令行中输入 jupyter notebook。
(3)在浏览器中打开 text_classification_with_hub.ipynb。
(4)逐步运行代码,查看运行结果。

11.2.3 代码解析

本次实战核心代码为 text_classification_with_hub.ipynb,下面进行详细的逐段解析。首先导入相关的函数库。

```
from __future__ import absolute_import, division, print_function, unicode_
literalsimport numpy as npimport tensorflow as tfimport tensorflow_hub as
hubimport tensorflow_datasets as tfdsprint("Version: ", tf.__version__)print("
Eager mode: ", tf.executing_eagerly())print("Hub version: ", hub.__version__)
print("GPU is", "available" if tf.config.experimental.list_physical_devices("
GPU") else "NOT AVAILABLE")
Version:  2.0.0
Eager mode:  True
Hub version:  0.6.0
GPU is available
```

下载 IMDB 数据集:IMDB 数据集可以在 TensorFlow 数据集处获取。以下代码将 IMDB 数据集下载至本地机器中。

```
#将训练集按照 6:4 的比例进行切割,从而最终将得到 15 000 个训练样本,10 000 个验证样本以及
#25 000 个测试样本
train_validation_split = tfds.Split.TRAIN.subsplit([6, 4])(train_data,
validation_data), test_data = tfds.load(name="imdb_reviews", split=(train_
validation_split, tfds.Split.TEST), as_supervised=True)
HBox(children=(IntProgress(value=0, description='Writing...', max=2500, style
=ProgressStyle(description_width=…
Dataset imdb_reviews downloaded and prepared to /home/kbuilder/tensorflow_
datasets/imdb_reviews/plain_text/0.1.0. Subsequent calls will reuse this data.
```

探索数据：下面了解一下数据的格式。每一个样本都是一个表示电影评论和相应标签的句子。该句子不以任何方式进行预处理。标签是一个值为 0 或 1 的整数，其中 0 代表消极评论，1 代表积极评论。

1. 构建模型

神经网络由堆叠的层来构建，这需要从 3 个主要方面来进行体系结构决策：如何表示文本？模型里有多少层？每个层里有多少隐藏单元（hiddenunits）？

本示例中，输入数据由句子组成。预测的标签为 0 或 1。表示文本的一种方式是将句子转换为嵌入向量（Embeddings Vectors）。可以使用一个预先训练好的文本嵌入（Text Embedding）作为首层，这将具有 3 个优点：不必担心文本预处理；可以从迁移学习中受益；嵌入具有固定长度，更易于处理。

针对此示例将使用 TensorFlow Hub 中名为 google/tf2-preview/gnews-swivel-20dim/1 的一种预训练文本嵌入（Text Embedding）模型。

为了达到本教程的目的，还有其他 3 种预训练模型可供测试。

（1） google/tf2-preview/gnews-swivel-20dim-with-oov/1——类似 google/tf2-preview/gnews-swivel-20dim/1，但 2.5% 的词汇转换为未登录词桶（OOV buckets）。如果任务的词汇与模型的词汇没有完全重叠，这将会有所帮助。

（2） google/tf2-preview/nnlm-en-dim50/1——一个拥有约 1M 词汇量且维度为 50 的更大的模型。

（3） google/tf2-preview/nnlm-en-dim128/1——拥有约 1M 词汇量且维度为 128 的更大的模型。

首先创建一个使用 TensorFlow Hub 模型嵌入（Embed）语句的 Keras 层，并在几个输入样本中进行尝试。请注意无论输入文本的长度如何，嵌入输出的形状都是（num_examples，embedding_dimension）。

```
embedding = "https://hub.tensorflow.google.cn/google/tf2-preview/gnews-swivel
-20dim/1"hub_layer = hub.KerasLayer(embedding, input_shape=[],  dtype=tf.
string, trainable=True)hub_layer(train_examples_batch[:3])
<tf.Tensor: id=402, shape=(3, 20), dtype=float32, numpy=
array([[ 3.9819887 , -4.4838037 ,  5.177359  , -2.3643482 , -3.2938678 ,
        -3.5364532 , -2.4786978 ,  2.5525482 ,  6.688532  , -2.3076782 ,
        -1.9807833 ,  1.1315885 , -3.0339816 , -0.7604128 , -5.743445  ,
         3.4242578 ,  4.790099  , -4.03061   , -5.992149  , -1.7297493 ],
       [ 3.4232912 , -4.230874  ,  4.1488533 , -0.29553518, -6.802391  ,
        -2.5163853 , -4.4002395 ,  1.905792  ,  4.7512794 , -0.40538004,
        -4.3401685 ,  1.0361497 ,  0.9744097 ,  0.71507156, -6.2657013 ,
         0.16533905,  4.560262  , -1.3106939 , -3.1121316 , -2.1338716 ],
       [ 3.8508697 , -5.003031  ,  4.8700504 , -0.04324996, -5.893603  ,
        -5.2983093 , -4.004676  ,  4.1236343 ,  6.267754  ,  0.11632943,
        -3.5934832 ,  0.8023905 ,  0.56146765,  0.9192484 , -7.3066816 ,
         2.8202746 ,  6.2000837 , -3.5709393 , -4.564525  , -2.305622  ]],
      dtype=float32)>
```

现在构建完整模型.

```
model = tf.keras.Sequential()model.add(hub_layer)model.add(tf.keras.layers.
Dense(16, activation='relu'))model.add(tf.keras.layers.Dense(1, activation='
sigmoid'))model.summary()
Model: "sequential"
_____
Layer (type)                 Output Shape              Param #
=================================================================
keras_layer (KerasLayer)     (None, 20)                400020
_____
dense (Dense)                (None, 16)                336
_____
dense_1 (Dense)              (None, 1)                 17
=================================================================
Total params: 400,373
Trainable params: 400,373
Non-trainable params: 0
_____
```

2. 层按顺序堆叠以构建分类器

第一层是 TensorFlow Hub 层。这一层使用一个预训练的保存好的模型来将句子映射为嵌入向量（Embedding Vector）。使用的预训练文本嵌入模型（google/tf2-preview/gnews-swivel-20dim/1）将句子切割为符号，嵌入每个符号然后进行合并。最终得到的维度是(num_examples，embedding_dimension)。

该定长输出向量通过一个有 16 个隐层单元的全连接层（Dense）进行管道传输。

最后一层与单个输出节点紧密相连。使用 Sigmoid 激活函数，其函数值为介于 0~1 的浮点数，表示概率或置信水平。

3. 编译模型、损失函数与优化器

一个模型需要损失函数和优化器来进行训练。由于这是一个二分类问题且模型输出概率值（一个使用 Sigmoid 激活函数的单一单元层），在此将使用 binary_crossentropy 损失函数。

这不是损失函数的唯一选择，例如，可以选择 mean_squared_error。但是，一般来说 binary_crossentropy 更适合处理概率——它能够度量概率分布之间的"距离"，或者在示例中，指的是度量 ground-truth 分布与预测值之间的"距离"。稍后，当研究回归问题（例如，预测房价）时，将介绍如何使用另一种叫作均方误差的损失函数。现在配置模型来使用优化器和损失函数。

```
model.compile(optimizer='adam', loss='binary_crossentropy', metrics=
['accuracy'])
```

11.2.4 训练测试

1. 训练模型

以 512 个样本的 mini-batch 大小迭代 20 个 epoch 来训练模型。这是指对 x_train 和 y_train 张量中所有样本的 20 次迭代。在训练过程中，监测来自验证集的 10 000 个样本上的损失值(loss)和准确率(accuracy)。

```
history=model.fit(train_data.shuffle(10000).batch(512),   epochs=20,
validation_data=validation_data.batch(512), verbose=1)
Epoch 1/20
30/30 [==============================] - 5s 153ms/step - loss: 0.9062 -
accuracy: 0.4985 - val_loss: 0.0000e+00 - val_accuracy: 0.0000e+00
Epoch 20/20
30/30 [==============================] - 3s 109ms/step - loss: 0.1899 -
accuracy: 0.9349 - val_loss: 0.2960 - val_accuracy: 0.8751
```

2. 评估模型

下面来看模型的表现如何。将返回两个值：损失值(loss)(一个表示误差的数字,值越低越好)与准确率(accuracy)。

```
results = model.evaluate(test_data.batch(512), verbose=2)for name, value in zip
(model.metrics_names, results):   print("%s: %.3f" % (name, value))
49/49 - 2s - loss: 0.3163 - accuracy: 0.8651
loss: 0.316
accuracy: 0.865
```

这种十分朴素的方法得到了约 86.5% 的准确率。

11.3 使用预训练的卷积神经网络绘制油画

你是否希望可以像毕加索或梵高一样绘画？本次实战使用深度学习来用其他图像的风格创造一个图像，这被称为神经风格迁移。

11.3.1 背景原理

神经风格迁移是一种优化技术，用于将两个图像(一个内容图像和一个风格参考图像,如著名画家的一个作品)混合在一起,使输出的图像看起来像源内容图像,但是用了参考图像的风格。

神经风格迁移的两大关键就是风格表征和图像重建。风格表征就是从图像中进行风格的提取和表示,建立风格的模型。而图像重建就是根据提取出的风格特征来重建一幅图像。

神经风格迁移可以让一幅图像拥有不同的风格,例如将照片转为油画风格、浮世绘风格等。如果想知道一幅场景由梵高来画会是怎么样的,那么只需要输入一张梵高的作品作为风格图像，神经风格迁移就可以学习图像里的风格并把它应用到另一张图像上去，这样也就能够知道梵高会怎样画这幅画了。

11.3.2 安装操作

本次实战既可以直接运行 python styletranfer.py，也可以安装 jupyter notebook。进入该项目目录下，输入命令 jupyter notebook 查看相关解析与运行代码 style_transfer.ipynb。

本节所用代码文件结构如表 11-3 所示。

表 11-3 代码文件结构

文 件 名 称	实 现 功 能	文 件 名 称	实 现 功 能
styletransfer.py	神经风格迁移 Python 版	style_transfer.ipynb	神经风格迁移 Jupyter Notebook 版

安装步骤如下。

(1) 进入 Anaconda Prompt 命令行，切换至本章根目录。

(2) 进入本节代码的目录。

(3) 在命令行中输入 python styletransfer.py 或进入 jupyter notebook 打开 style_transfer.ipynb。

(4) 运行代码，查看运行结果。

11.3.3 代码解析

本次实战核心代码为 style_transfer.ipynb。下面是分段步骤与解析。

(1) 导入和配置模块。

```
import tensorflow as tf
import IPython.display as display
import matplotlib.pyplot as plt
import matplotlib as mpl
mpl.rcParams['figure.figsize'] = (12,12)
mpl.rcParams['axes.grid'] = False
import numpy as np
import PIL.Image
import time
import functools
```

(2) 下载图像并选择风格图像和内容图像。

```
content_path = tf.keras.utils.get_file('YellowLabradorLooking_new.jpg',
' https://storage. googleapis. com/download. tensorflow. org/example _ images/
YellowLabradorLooking_new.jpg')
Downloading data from https://storage.googleapis.com/download.tensorflow.org/
example_images/Green_Sea_Turtle_grazing_seagrass.jpg
32768/29042 [==============================] - 0s 0us/step
Downloading data from https://storage.googleapis.com/download.tensorflow.org/
```

```
example_images/Vassily_Kandinsky%2C_1913_-_Composition_7.jpg
196608/195196 [==============================] - 0s 0us/step
```

(3) 将输入可视化

定义一个加载图像的函数,并将其最大尺寸限制为 512 像素。

```
def load_img(path_to_img):
  max_dim = 512
  img = tf.io.read_file(path_to_img)
  img = tf.image.decode_image(img, channels=3)
  img = tf.image.convert_image_dtype(img, tf.float32)
  shape = tf.cast(tf.shape(img)[:-1], tf.float32)
  long_dim = max(shape)
  scale = max_dim / long_dim
  new_shape = tf.cast(shape * scale, tf.int32)
  img = tf.image.resize(img, new_shape)
  img = img[tf.newaxis, :]
  return img
```

(4) 创建一个简单的函数来显示图像。

```
def imshow(image, title=None):
  if len(image.shape) > 3:
    image = tf.squeeze(image, axis=0)
  plt.imshow(image)
  if title:
    plt.title(title)
```

创意生成图像对比如图 11-2 所示。

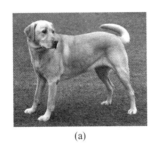

(a)　　　　　　　　　　　(b)

图 11-2　创意生成图像对比

(5) 定义内容和风格的表示。

使用模型的中间层来获取图像的内容和风格表示。从网络的输入层开始,前几个层的激励响应表示边缘和纹理等低级特征。随着层数加深,最后几层代表更高级的特征——实体部分,如轮子或眼睛。

在本次实战中,使用的是 VGG19 网络结构,这是一个已经预训练好的图像分类网络。这些中间层是从图像中定义内容和风格的表示所必需的。对于一个输入图像,尝试匹配这些中间层的相应风格和内容目标的表示。

加载 VGG19 并在图像上测试它以确保正常运行。

```
x = tf.keras.applications.vgg19.preprocess_input(content_image * 255)
x = tf.image.resize(x, (224, 224))
vgg = tf.keras.applications.VGG19(include_top=True, weights='imagenet')
prediction_probabilities = vgg(x)
prediction_probabilities.shape
Downloading data from https://github.com/fchollet/deep-learning-models/
releases/download/v0.1/vgg19_weights_tf_dim_ordering_tf_kernels.h5
574717952/574710816 [==============================] - 34s 0us/step
TensorShape([1, 1000])
predicted_top_5 = tf.keras.applications.vgg19.decode_predictions(prediction_
probabilities.numpy())[0]
[(class_name, prob) for (number, class_name, prob) in predicted_top_5]
Downloading data from https://storage.googleapis.com/download.tensorflow.org/
data/imagenet_class_index.json
40960/35363 [==============================] - 0s 0us/step
[('loggerhead', 0.74297667),
 ('leatherback_turtle', 0.11357909),
 ('hermit_crab', 0.054411974),
 ('terrapin', 0.039235227),
 ('mud_turtle', 0.012614701)]
```

现在,加载没有分类部分的 VGG19,并列出各层的名称。

```
vgg = tf.keras.applications.VGG19(include_top=False, weights='imagenet')
for layer in vgg.layers:
  print(layer.name)
Downloading data from https://github.com/fchollet/deep-learning-models/
releases/download/v0.1/vgg19_weights_tf_dim_ordering_tf_kernels_notop.h5
80142336/80134624 [==============================] - 6s 0us/step
```

从网络中选择中间层的输出以表示图像的风格和内容。

```
#内容层将提取出特征图
content_layers = ['block5_conv2']
#风格层
style_layers = ['block1_conv1','block2_conv1','block3_conv1', 'block4_conv1',
'block5_conv1']
num_content_layers = len(content_layers)
num_style_layers = len(style_layers)
```

对于用于表示风格和内容的中间层,为什么预训练的图像分类网络中的这些中间层的输出允许定义风格和内容的表示?从高层理解,为了使网络能够实现图像分类(该网络已被训练过),它必须理解图像。这需要将原始图像作为输入像素并构建内部表示,这个内部表示将原始图像像素转换为对图像中存在的特征的复杂理解。

这也是卷积神经网络能够很好推广的一个原因:它们能够捕获不变性并定义类别(例

如猫与狗）之间的特征，这些特征与背景噪声和其他干扰无关。因此，将原始图像传递到模型输入和分类标签输出之间的某处的这一过程，可以视作复杂的特征提取器。通过这些模型的中间层，就可以描述输入图像的内容和风格。

（6）建立模型。

使用 tf.keras.applications 中的网络可以非常方便地利用 Keras 的功能接口提取中间层的值。在使用功能接口定义模型时，需要指定输入和输出。

```
model = Model(inputs, outputs)
```

以下函数构建了一个 VGG19 模型，该模型返回一个中间层输出的列表。

```
def vgg_layers(layer_names):
  """ Creates a vgg model that returns a list of intermediate output values."""
  #加载模型与预训练 VGG 网络，并在 ImageNet 数据集上训练
  vgg = tf.keras.applications.VGG19(include_top=False, weights='imagenet')
  vgg.trainable = False
  outputs = [vgg.get_layer(name).output for name in layer_names]
  model = tf.keras.Model([vgg.input], outputs)
  return model
```

然后建立模型。

```
style_extractor = vgg_layers(style_layers)
style_outputs = style_extractor(style_image * 255)
#查看每层输出的统计信息
for name, output in zip(style_layers, style_outputs):
  print(name)
  print("  shape: ", output.numpy().shape)
  print("  min: ", output.numpy().min())
  print("  max: ", output.numpy().max())
  print("  mean: ", output.numpy().mean())

block1_conv1
  shape:  (1, 336, 512, 64)
  min:  0.0
  max:  835.5256
  mean:  33.97525
block2_conv1
  shape:  (1, 168, 256, 128)
  min:  0.0
  max:  4625.8857
  mean:  199.82687
block3_conv1
  shape:  (1, 84, 128, 256)
  min:  0.0
  max:  8789.239
  mean:  230.78099
block4_conv1
```

```
    shape:  (1, 42, 64, 512)
    min:  0.0
    max:  21566.135
    mean:  791.24005
block5_conv1
    shape:  (1, 21, 32, 512)
    min:  0.0
    max:  3189.2542
    mean:  59.179478
```

（7）风格计算

图像的内容由中间特征图的值表示。

事实证明，图像的风格可以通过不同特征图上的平均值和相关性来描述。通过在每个位置计算特征向量的外积，并在所有位置对该外积进行平均，可以计算出包含此信息的 Gram 矩阵。

这可以使用 tf.linalg.einsum 函数来实现。

```
def gram_matrix(input_tensor):
    result = tf.linalg.einsum('bijc,bijd->bcd', input_tensor, input_tensor)
    input_shape = tf.shape(input_tensor)
    num_locations = tf.cast(input_shape[1] * input_shape[2], tf.float32)
    return result/(num_locations)
```

提取风格和内容，构建一个返回风格和内容张量的模型。

```
class StyleContentModel(tf.keras.models.Model):
  def __init__(self, style_layers, content_layers):
    super(StyleContentModel, self).__init__()
    self.vgg =  vgg_layers(style_layers + content_layers)
    self.style_layers = style_layers
    self.content_layers = content_layers
    self.num_style_layers = len(style_layers)
    self.vgg.trainable = False
  def call(self, inputs):
    "Expects float input in [0,1]"
    inputs = inputs * 255.0
    preprocessed_input = tf.keras.applications.vgg19.preprocess_input(inputs)
    outputs = self.vgg(preprocessed_input)
    style_outputs, content_outputs = (outputs[:self.num_style_layers],
                                     outputs[self.num_style_layers:])
    style_outputs = [gram_matrix(style_output)
                     for style_output in style_outputs]
    content_dict = {content_name:value
                    for content_name, value
                    in zip(self.content_layers, content_outputs)}
```

```
            style_dict = {style_name:value
                          for style_name, value
                          in zip(self.style_layers, style_outputs)}
        return {'content':content_dict, 'style':style_dict}
```

在图像上调用此模型，可以返回 style_layers 的 gram 矩阵（风格）和 content_layers 的内容。

```
extractor = StyleContentModel(style_layers, content_layers)
results = extractor(tf.constant(content_image))
style_results = results['style']

print('Styles:')
for name, output in sorted(results['style'].items()):
    print("  ", name)
    print("    shape: ", output.numpy().shape)
    print("    min: ", output.numpy().min())
    print("    max: ", output.numpy().max())
    print("    mean: ", output.numpy().mean())

print("Contents:")
for name, output in sorted(results['content'].items()):
    print("  ", name)
    print("    shape: ", output.numpy().shape)
    print("    min: ", output.numpy().min())
    print("    max: ", output.numpy().max())
    print("    mean: ", output.numpy().mean())
```

梯度下降：使用此风格和内容提取器，现在可以实现风格传输算法。通过计算每个图像的输出和目标的均方误差来做到这一点，然后取这些损失值的加权和。

设置风格和内容的目标值。

```
style_targets = extractor(style_image)['style']
content_targets = extractor(content_image)['content']
```

定义一个 tf.Variable 来表示要优化的图像。为了快速实现这一点，使用内容图像对其进行初始化（tf.Variable 必须与内容图像的形状相同）。

```
image = tf.Variable(content_image)
```

由于这是一个浮点图像，因此定义一个函数来保持像素值在 0~1。

```
def clip_0_1(image):
    return tf.clip_by_value(image, clip_value_min=0.0, clip_value_max=1.0)
```

创建一个 optimizer。本实战推荐 LBFGS，但 Adam 也可以正常工作。

```
opt = tf.optimizers.Adam(learning_rate=0.02, beta_1=0.99, epsilon=1e-1)
```

为了优化它，使用两个损失的加权组合来获得总损失。

```
def style_content_loss(outputs):
    style_outputs = outputs['style']
    content_outputs = outputs['content']
    style_loss = tf.add_n([tf.reduce_mean((style_outputs[name]-style_targets
[name])**2) for name in style_outputs.keys()])
    style_loss *= style_weight / num_style_layers

    content_loss = tf.add_n([tf.reduce_mean((content_outputs[name]-content_
targets[name])**2) for name in content_outputs.keys()])
    content_loss *= content_weight / num_content_layers
    loss = style_loss + content_loss
    return loss
```

使用 tf.GradientTape 来更新图像。

```
@tf.function()
def train_step(image):
  with tf.GradientTape() as tape:
    outputs = extractor(image)
    loss = style_content_loss(outputs)

  grad = tape.gradient(loss, image)
  opt.apply_gradients([(grad, image)])
  image.assign(clip_0_1(image))
```

现在,运行几步来测试一下,测试的初步渲染图像如图 11-3 所示。

```
train_step(image)
train_step(image)
train_step(image)
tensor_to_image(image)
```

图 11-3　初步渲染图像

运行正常,来执行一个更长的优化。

```
start = time.time()
epochs = 10
steps_per_epoch = 100
step = 0
for n in range(epochs):
  for m in range(steps_per_epoch):
    step += 1
    train_step(image)
    print(".", end='')
  display.clear_output(wait=True)
  display.display(tensor_to_image(image))
  print("Train step: {}".format(step))
end = time.time()
print("Total time: {:.1f}".format(end-start))
```

11.3.4 训练测试

运行上述程序后得到的最终风格样式迁移结果如图 11-4 和图 11-7 所示，通过提取不同频率的颜色与边缘特征，如图 11-5 和图 11-6 所示，对油画的整体样式风格进行特征学习后再对新的图像进行样式迁移，最终得到了全新的图像训练结果。

图 11-4　仿制油画效果图

```
Total time: 21.5
```

总变分损失：此实现只是一个基础版本，它的一个缺点是它会产生大量的高频误差。可以直接通过正则化图像的高频分量来减少这些高频误差。在风格转移中，这通常被称为总变分损失。

```
def high_pass_x_y(image):
  x_var = image[:,:,1:,:] - image[:,:,:-1,:]
  y_var = image[:,1:,:,:] - image[:,:-1,:,:]
  return x_var, y_var
x_deltas, y_deltas = high_pass_x_y(content_image)
```

```
plt.figure(figsize=(14,10))
plt.subplot(2,2,1)
imshow(clip_0_1(2*y_deltas+0.5), "Horizontal Deltas: Original")
plt.subplot(2,2,2)
imshow(clip_0_1(2*x_deltas+0.5), "Vertical Deltas: Original")
x_deltas, y_deltas = high_pass_x_y(image)
plt.subplot(2,2,3)
imshow(clip_0_1(2*y_deltas+0.5), "Horizontal Deltas: Styled")
plt.subplot(2,2,4)
imshow(clip_0_1(2*x_deltas+0.5), "Vertical Deltas: Styled")
```

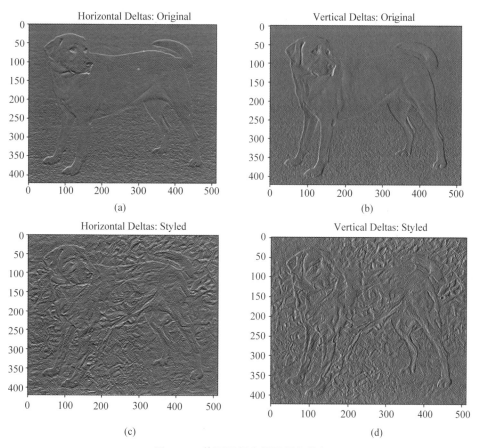

图 11-5 从不同频率提取图像特征

这显示了高频分量如何增加。而且，本质上高频分量是一个边缘检测器。可以从 Sobel 边缘检测器获得类似的输出，例如：

```
plt.figure(figsize=(14,10))

sobel = tf.image.sobel_edges(content_image)
plt.subplot(1,2,1)
imshow(clip_0_1(sobel[...,0]/4+0.5), "Horizontal Sobel-edges")
```

```
plt.subplot(1,2,2)
imshow(clip_0_1(sobel[...,1]/4+0.5), "Vertical Sobel-edges")
```

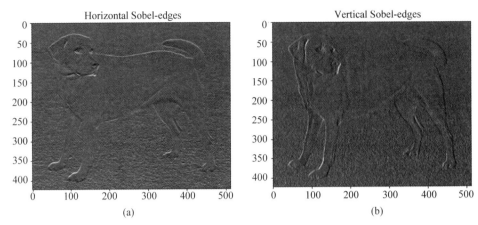

图 11-6　水平与垂直两个维度提取图像不同频率特征效果图

与此相关的正则化损失是这些值的平方和。

```
def total_variation_loss(image):
  x_deltas, y_deltas = high_pass_x_y(image)
  return tf.reduce_sum(tf.abs(x_deltas)) + tf.reduce_sum(tf.abs(y_deltas))
```

重新进行优化，选择 total_variation_loss 的权重。

```
total_variation_loss(image).numpy()
```

现在，将它加入 train_step 函数中。

```
@tf.function()
def train_step(image):
  with tf.GradientTape() as tape:
    outputs = extractor(image)
    loss = style_content_loss(outputs)
    loss += total_variation_weight * tf.image.total_variation(image)

  grad = tape.gradient(loss, image)
  opt.apply_gradients([(grad, image)])
  image.assign(clip_0_1(image))
```

重新初始化优化的变量。

```
image = tf.Variable(content_image)
```

进行优化。

```
import time
start = time.time()
epochs = 10
```

```
steps_per_epoch = 100
step = 0
for n in range(epochs):
  for m in range(steps_per_epoch):
    step += 1
    train_step(image)
    print(".", end='')
  display.clear_output(wait=True)
  display.display(tensor_to_image(image))
  print("Train step: {}".format(step))
end = time.time()
print("Total time: {:.1f}".format(end-start))
Total time: 22.8
```

最后,保存结果。

```
file_name = 'stylized-image1.png'
tensor_to_image(image).save(file_name)
```

图 11-7 就是最终优化后的图像,由照片风格向着油画效果转换的结果。

图 11-7　最终优化结果图像展示

结 语

本书通过全面介绍基于深度学习的人工智能相关领域，引导大家通过实践来体会其背后的原理，提高计算机视觉、自然语言处理、强化学习等热门领域的动手能力和深度思考。通过有限的案例引导大家对于人工智能的未来产生无限思考。

当前，人工智能技术在帮助人类开展信息收集、信息分析以及开展决策的过程中发挥着日益重要的作用。在很多领域中，人工智能技术的信息处理能力已经远超人类脑力，在此背景下，人工智能能够替代人类完成很多复杂的工作。毋庸置疑的是，在人工智能的发展中，高度的智能化是其主要的发展趋势，而可以预见的"高度智能化"，则体现为深度学习能力的提升，即人工智能技术不仅能够替代人类完成一些较为复杂的体力劳动，而且能够具备独立思考与独立分析的能力。具体而言，从未来人工智能技术的发展基础来看，一方面，大数据技术能够为人工智能开展深度学习带来更为丰富的素材，因此，大数据技术能够在人工智能技术发展中发挥出不容忽视的推动作用；另一方面，云计算、GPU等，是人工智能具备独立思考与独立分析能力的重要支撑，相对于大数据在人工智能技术发展中的作用而言，云计算、GPU等更像是人工智能技术中的"消化"系统、因此，云计算、GPU技术的发展，为人工智能技术的发展带来了难得的契机。

人工智能发展趋势将呈现出3种基本特征。首先，基于深度学习的人工智能技术，将呈现出更快的发展速度。相对于以人为开发主体的技术发展模式而言，基于深度学习的人工智能技术能够对当前社会中存在的知识和经验进行吸收，虽然人类在获取这些知识与经验的过程中经历了漫长的发展历史，但是对于人工智能而言，这一知识与经验吸收过程所占用的时间将会十分短暂。而对这些知识和经验进行吸收的结果则体现为人工智能对自身的持续完善。其次，在深度学习基础上，人工智能将能够展现出更加强大的信息挖掘能力与人机交互能力。近年来，人工智能的概念和技术在逐渐向各个领域中渗透，一些依托人工智能技术所开发出的衍生品也已经初步具备了良好的人机交互能力，这种人机交互能力促使这些衍生品的用户获得了更为良好的产品使用体验。随着人工智能的高度智能化特别是深度学习能力的提升，这些基于人工智能技术所生产的衍生品也将呈现出更为强大的人机交互能力，从而为社会大众的生活以及各行各业的生产带来更多便利。在过去40年里，超算帮助人们解决了从宇宙天体到蛋白质分子一系列非常复杂的模拟问题。如今，超算面向生物医药研制、人工智能医疗领域的成效越来越凸显。以全基因组信息关联性分析为例，之前需要几年的时间进行分析，采用超算后可以将时间缩短为十小时。而在基因测序分析选择靶向药物用于靶向治疗上，则可以将时间缩短到30分钟。随着大数据、人工智能对超算的需求越来越强劲，可以说人工智能让超算从高大上的科学领域扩展到了和人们生活息息相关的应用领域。高性能计算本身为人工智能的崛起提供了一个新的计算引擎，而大数据和人工智能也牵引着高性能计算呈现出很多新的形态。人工智能需要超算作为后台进行越来越强大的模型训练、智能推理、关联分析,导致超算除了用于科学工程计算领域,应用范围也越来越多元化。今后,除传统的大气、海洋、能源 CFD 的高性能计算应用领域外,人工智能带来的超算应用会越来越多。两者可以互相影响,相辅相成。例如,在广州超算中心,原来做超算和做人工智能的两个团队之前交互不多,随着近几年 GPU 出现以后交互开始增多。很多超算中的经验可以被人工智能的算法和应用借鉴,例如稀疏矩阵、数据库等已经在超算中持续优化多年的领域,可以让人工智能使用者更方便地使用。

  本书到这里就结束了,希望大家能通过深度学习的实战案例,结合未来人工智能发展趋势,为工程界和学术界作出更大贡献。

参考文献

[1]   https://www.anaconda.com.
[2]   https://www.tensorflow.org/? hl=zh-cn.
[3]   https://developer.nvidia.com/zh-cn/cuda-toolkit.
[4]   https://jupyter.org/.
[5]   https://www.tensorflow.org/? hl=zh-cn.
[6]   https://archive.ics.uci.edu/ml/datasets/Heart+Disease.
[7]   https://archive.ics.uci.edu/ml/datasets/auto+mpg.
[8]   http://archive.ics.uci.edu/ml/datasets/higgs.
[9]   https://www.tensorflow.org/tutorials/images/data_augmentation? hl=zh-cn.
[10] SIMONYAN K, ZISSERMAN A. Very deep convolutional networks for large-scale image recognition[J]. arXiv, 2014, 09(1556): 1-14.
[11] KRIZHEVSKY A, HINTON G. Learning multiple layers of features from tiny images[J]. Citeseer, 2009, 07(1): 1-58.
[12] Single image super-resolution with deep neural networks [EB/OL], 2019. https://krasserm.github.io/2019/09/04/super-resolution.
[13] LIM B, SON S, KIM H, et al. Enhanced deep residual networks for single image super-resolution [C]//Proceedings of the IEEE conference on computer vision and pattern recognition workshops. 2017: 136-144.
[14] YU J, FAN Y, YANG J, et al. Wide activation for efficient and accurate image super-resolution[J]. arXiv, 2018, 08(8718): 1-8.
[15] LEDIG C, THEIS L, HUSZÁR F, et al. Photo-realistic single image super-resolution using a generative adversarial network[C]//Proceedings of the IEEE conference on computer vision and pattern recognition. 2017: 4681-4690.
[16] HOCHREITER S, SCHMIDHUBER J. Long short-term memory[J]. Neural computation, 1997, 9(8): 1735-1780.
[17] https://cocodataset.org/#home.
[18] SZEGEDY C, VANHOUCKE V, IOFFE S, et al. Rethinking the inception architecture for computer vision [C]//Proceedings of the IEEE conference on computer vision and pattern recognition. 2016: 2818-2826.
[19] Yuthon'Blog. Notes: From Faster R-CNN to Mask R-CNN [EB/OL], 2017. https://www.yuthon.com/post/tutorials/notes-from-faster-r-cnn-to-mask-r-cnn/#yesterday-background-and-pre-works-of-mask-r-cnn.
[20] GIRSHICK R, DONAHUE J, DARRELL T, et al. Rich feature hierarchies for accurate object detection and semantic segmentation[C]//Proceedings of the IEEE conference on computer vision and pattern recognition. 2014: 580-587.
[21] REN S, HE K, GIRSHICK R, et al. Faster r-cnn: Towards real-time object detection with region proposal networks[J]. Advances in neural information processing systems, 2015, 28: 91-99.
[22] ZHANG K, ZHANG Z, LI Z, et al. Joint face detection and alignment using multitask cascaded convolutional networks[J]. IEEE Signal Processing Letters, 2016, 23(10): 1499-1503.

[23] GIRSHICK R. Fast R-CNN[C]. 2015 IEEE International Conference on Computer Vision, 2015, 169: 1440-1448.

[24] HUANG S, RAMANAN D. Expecting the Unexpected: Training Detectors for Unusual Pedestrians with Adversarial Imposters[J]. IEEE, 2017.

[25] SCHROFF F, KALENICHENKO D, PHILBIN J. FaceNet: A Unified Embedding for Face Recognition and Clustering[J]. IEEE, 2015.

[26] REDMON J, DIVVALA S, GIRSHICK R, et al. You Only Look Once: Unified, Real-Time Object Detection[C]// Computer Vision & Pattern Recognition. IEEE, 2016.

[27] REDMON J, FARHADI A. YOLO9000: Better, Faster, Stronger[C]// IEEE Conference on Computer Vision & Pattern Recognition. IEEE, 2017: 6517-6525.

[28] FARHADI A, REDMON J. Yolov3: An incremental improvement[C]//Computer Vision and Pattern Recognition. 2018: 1804.02767.

[29] BOCHKOVSKIY A, WANG C Y, LIAO H. YOLOv4: Optimal Speed and Accuracy of Object Detection[J].arXiv, 2020, 04(10934): 1-14.

[30] LIU W, ANGUELOV D, ERHAN D, et al. SSD: Single Shot MultiBox Detector[J]. Springer, Cham, 2016.

[31] https://github.com/thtrieu/darkflow.

[32] LIN T Y, GOYAL P, GIRSHICK R, et al. Focal loss for dense object detection[C]//Proceedings of the IEEE international conference on computer vision. 2017: 2980-2988.

[33] LIN T Y, GOYAL P, GIRSHICK R, et al. Focal Loss for Dense Object Detection[J]. IEEE Transactions on Pattern Analysis & Machine Intelligence, 2017, (99): 2999-3007.

[34] HE K, ZHANG X, REN S, et al. Deep residual learning for image recognition[C]//Proceedings of the IEEE conference on computer vision and pattern recognition. 2016: 770-778.

[35] HOWARD A G, ZHU M, CHEN B, et al. MobileNets: Efficient Convolutional Neural Networks for Mobile Vision Applications[J]. arXiv, 2017, 04(4861): 1-8.

[36] LECUN Y, BOTTOU L, BENGIO Y, et al. Gradient-based learning applied to document recognition[J]. Proceedings of the IEEE, 1998, 86(11): 2278-2324.

[37] BOURLARD H, KAMP Y. Auto-association by multilayer perceptrons and singular value decomposition[J]. Biological cybernetics, 1988, 59(4): 291-294.

[38] HINTON G E, SALAKHUTDINOV R R. Reducing the dimensionality of data with neural networks[J]. science, 2006, 313(5786): 504-507.

[39] KIPF T N, WELLING M. Variational graph auto-encoders[J]. NIPS, 2016.

[40] CHENG-ZHI ANAN HUANG, ASHISH V, JAKOB. Music Transformer: Generating Music with Long-Term Structure[J]. arXiv, 2018,09(4281): 1-10.

[41] COLAH'S BLOG. Understanding LSTM Networks[EB/OL], 2015. http://colah.github.io/posts/2015-08-Understanding-LSTMs/.

[42] SOCHER R, PERELYGIN A, WU J Y, et al. Recursive Deep Models for Semantic Compositionality Over a Sentiment Treebank[C]. In: Proceedings of the 2013 Conference on Empiricalin Natural Language Processing, 2013(01): 1631-1642.

[43] DAN W. Chinese to English automatic patent machine translation at SIPO[J]. World Patent Information, 2009, 31(2): 137-139.

[44] ISOLA P, ZHU J Y, ZHOU T, et al. Image-to-Image Translation with Conditional Adversarial Networks[C]// IEEE Conference on Computer Vision & Pattern Recognition. IEEE, 2016.

[45] ZHU J Y, PARK T, ISOLA P, et al. Unpaired Image-to-Image Translation using Cycle-Consistent Adversarial Networks[J]. IEEE, 2017.

[46] MNIH V, KAVUKCUOGLU K, SILVER D, et al. Playing Atari with Deep Reinforcement Learning[J]. Computer Science, 2013.

[47] CHEN K. Deep Reinforcement Learning for Flappy Bird[J]. Stanford University, 2015, 09(01): 1-1.2

[48] SILVER D, HUBERT T, SCHRITTWIESER J, et al. Mastering Chess and Shogi by Self-Play with a General Reinforcement Learning Algorithm[J]. arXiv, 2017, 12(1815): 1-13.

[49] SILVER D, SCHRITTWIESER J, SIMONYAN K, et al. Mastering the game of Go without human knowledge[J]. Nature, 2017, 550(7676): 354-359.

[50] VOLODYMYR M, KORAY K, DAVID S, et al. Human-level control through deep reinforcement learning[J]. Nature, 2019, 518(7540): 529-33.

[51] https://blog.csdn.net/weixin_35770067/.

[52] 赵英俊. 走向TensorFlow 2.0：深度学习应用编程快速入门[M]. 北京：电子工业出版社, 2019.

[53] 程世东. 深度学习私房菜：跟着案例学TensorFlow 2.0[M]. 北京：电子工业出版社, 2019.

[54] 邓劲生, 庄春华. 实战深度学习——原理、框架及应用[M]. 北京：清华大学出版社, 2021.